QUANTUM PHYSICS FOR BEGINNERS

LEARN THE MAIN QUANTUM PHYSICS PRINCIPLES,
USE THE POWER OF THE MIND,
UNDERSTAND THE UNIVERSE AND CHANGE YOUR
POINT OF VIEW ON LIFE

By: Ethan Walker

Table of Contents

Introduction

As the world of physics continues to develop, so does the public's knowledge of quantum physics. A new generation of physicists is replacing their chalkboards with calculators while others try to adjust to the new findings.

The most recent development in quantum physics is the non-locality of "entanglement." Entanglement is a pairing of particles that are correlated in such a way that the position of one particle depends on the position of the other particle. Physicists have observed this phenomenon in so-called "quantum states", where many particles are entangled. In these quantum states, it is possible for one entangled particle to be in one place and another entangled particle to be in another place. In these situations, one particle can influence the other without any distance between them.

As intriguing as this concept is, it is difficult for laymen to grasp since we cannot actually contain both particles in a single location at the same time. We can, however, simulate this phenomenon using a computer simulation in which entangled particles are located next to each other on a grid. Since these particles all reside on the same grid, they remain strongly correlated even if any two of them change position. This means that even though their positions are changed, their correlations do not go away because they automatically adjust themselves to each other's changes.

The science of quantum mechanics is one of the most powerful branches of modern physics. But why should you care? Quantum mechanics gives us a different way to see reality. Our conventional view of the universe involves measuring things and assigning them numbers or categorizing them into groups. This is all well and good, but it has some problems: not everything is always quantifiable, for one, and we often can't see what something is made of.

Quantum physics, on the other hand, explains what happens when we measure or observe something, and it gives us a more complete understanding of how the universe works. It tells us what things are made of and how they work. Furthermore, it predicts that some things can act in ways that defy common sense, as well as various observations made by scientists today.

Everyone knows about electrons, protons, neutrons, and such, but let's talk about quarks for a bit as well. Quarks are subatomic particles that make up matter in our universe. They're different from electrons and protons in that they're fermions—that is to say, they have half-integer values of charge, spin, and elementary charge (Lambda).

Similarly, neutrinos are subatomic particles that interact extremely weakly with other matter (and thus aren't detected). Unlike quarks, however, neutrinos do have mass—in fact, they have "lepton" mass (they aren't quite quarks because they're actually a type of lepton), which means they do interact with other matter at large distances. This means that neutrinos can be used to explore the universe at very high energies without interfering with other forms of matter (thereby casting fresh light on the origin and nature of dark matter).

In 1810, John Michell and Pierre Laplace published a paper proposing that light was particles (photon) and that the energy of these particles was dependent on their wavelength. The discovery of the photoelectric effect and the Einstein photoelectric equation would eventually demonstrate that light was wave-like in nature, which changed our understanding of light not only to waves but to particles (quanta). In 1911, Albert Einstein provided further evidence to support his hypothesis when he published papers on the photoelectric effect and on the field equations for special relativity.

A century later, there are still many questions left unanswered when considering what does and does not constitute matter. One of the most basic questions is, "What is a particle?" There have been numerous attempts to answer this question, but none of them have been entirely satisfactory. Within the past few years, physicists have taken a different approach to answer this seemingly simple question. Instead of defining particles as something that exhibits wave-like behavior (light), they have defined them as something that exhibits particle-like behavior (quanta). While this may seem like a small step forward, this new definition of particles is monumental in its implications. Because the new definition of particles allows us to better understand itself, it will allow us to better understand how nature operates. This understanding is key to physicists who hope for creating other types of particles. These other kinds of particles will be vital if they wish to understand dark matter and dark energy, thus affording us an understanding of what makes up the universe. The new view of quanta opens up new possibilities for particle physics research as well as science in general.

Chapter 1

Differences Between Classical Physics and Quantum Physics

Physicists are trying to understand the world - but to date, there has surfaced no proven, palpable theory to bring the two worlds together and finally help us understand where we come from, where we are, and where we are going (because, at the end of the day, these are the fundamental questions both classical and quantum physics propose).

In classical physics (as drawn out by Einstein's general relativity principle), the reality is made out of 4 dimensions (also called the space-time continuum). In this paradigm, gravitational fields are continuous entities.

In quantum mechanics, however, fields are not continuous but discontinuous. They are not defined by the 4 dimensions but by "quanta." As such, concepts like the "gravitational field" are missing from the world of quantum physics, which is also the biggest bridge classical physicists and quantum researchers have to build between their points of view.

This is not just a matter of fancy definitions. The world of quantum mechanics and the world of classical physics are incompatible because they describe reality in completely different ways, in dissimilar terms, and different perspectives that do not meet at any point.

In classical physics, things happen for a reason. They happen according to the old cause-and-effect dictum.

Nothing happens randomly, but because there is something else before it that has caused it.

In quantum physics, scientists do not see reality in terms of cause and effect, but in terms of particles jumping from one state to another based on probability rather than definite outcomes.

Why is reconciliation important, then, especially given that these two disciplines seem so different and at such a deep level?

Because reconciliation would create a whole theory that explains the universe on a small and large scale, where classical physics fails to give explanations of the microcosms, quantum physics would succeed. And where quantum physics fails to make sense when it is blown up to macro objects (remember the cat that was both dead and alive?), classical physics would be able to breathe in some logic.

Fundamentals About Classical Physics

Classical physics is invariant for time-reversal, and we have seen that this gives us serious problems when we try to find an explanation for the thermodynamic arrow of time. It is therefore important to investigate now on dynamic reversibility in quantum mechanics.

Before we have a look at particle physics, quantum theory, and the cosmos, we need a brief introduction to the concepts of classical physics:

- Energy
- Weight and mass
- Matter—solids, and liquids
- Measures and units

Energy

Energy must be transferred to an object to perform work on or heat it.

Newton's law of the conservation of energy states that it may be transformed from one form to another. **It cannot be created or destroyed.**

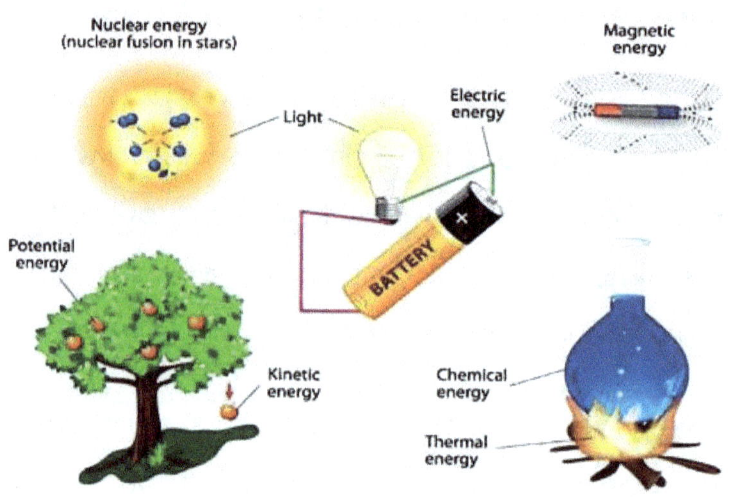

Forms of energy

Types of energy include:

- Kinetic energy (movement)
- Chemical energy (e.g. coal, natural gas, etc.)
- Thermal energy
- Magnetic energy

- Light energy

- Electric energy

- Gravitational potential energy

- Nuclear energy

Weight and Mass

Weight is, therefore, dependent on the gravitational force where the object is situated. According to the International System of Units (SI Units), weight is measured in Newtons with the symbol N.

Mass can be described simply as the amount of matter there is in an object. This measurement is given in kilograms or grams and is calculated by multiplying the object's volume by its density. An object's mass will be the same wherever it is measured because the object will always contain the same number of protons (amount of matter).

Matter

There are several forms of matter that we know of, but only three are of relevance to this book: solids, liquids, and gases.

- Solids. Solid objects have a defined shape because their atoms are packed tightly together (i.e., they have a high density). The atoms cannot move around and cannot be compressed into a smaller volume.

- Liquids. In liquids, the atoms are not so tightly packed, so they can flow around each other. Most liquids can be compressed into a smaller volume in a container, i.e. their atoms are forced closer together (they become denser).

- Gases. The atoms in gases are in constant movement and have a relatively large space between them.

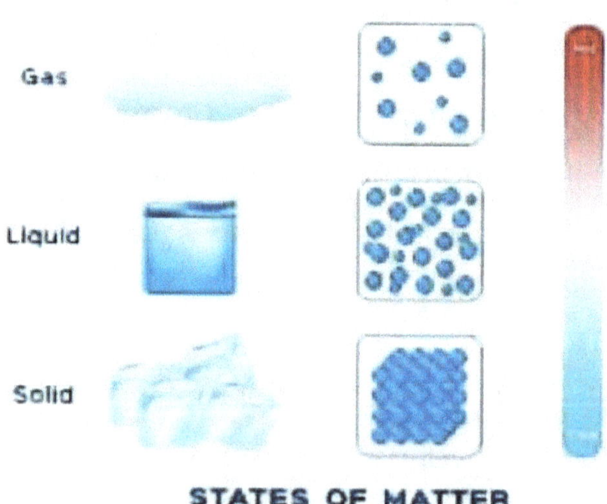

STATES OF MATTER

Measures and Units

- Density. The density of matter is a measure of how closely the atoms are packed together. Density is measured in kilograms per cubic meter and can be calculated by dividing an object's mass by its volume.

- Volume The amount of space that matter occupies. There are many ways of measuring volume, depending on whether you are measuring solids, liquids, or gases. The formulae for measuring different shapes of solids or containers of liquids or gases are different.

As I promised in the introduction, not to burden you with mathematics, other than E=mc2, I will leave you to find these formulae for yourself in further reading if you are interested. There are, however, some easy ways to measure small amounts of gases and liquids (see Fig 1).

Figure 1 shows the gas volume is measured by water displacement

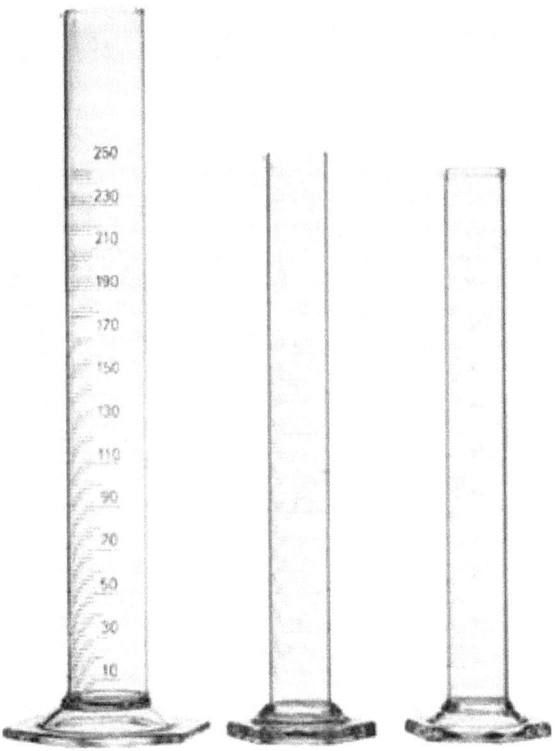

The units in which volume is measured can be confusing. The cuslomary system in the US differs from the imperial system in the UK, and both of those differ from the metric system. In 1960, an international system of units was introduced. Le Systeme Internationale Duties, now known simply as SI Units, are used by the scientific community to avoid confusion.

Fundamentals of Quantum Physics

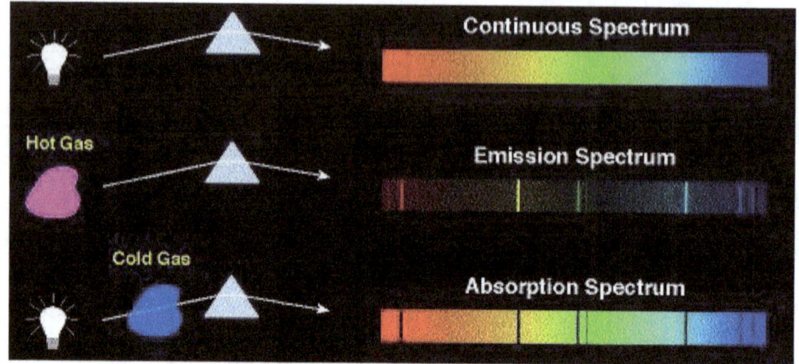

The Quantization of Light

This was a forward step taken by Albert Einstein in the year 1905. With him, he suggested that quantization did not just involve mathematical tricks. On it, he added that it also involved the beam of light energy that is in the individual packets- currently referred to as photons. Thus, the energy of a single photon will be given by the product of the frequency of the energy and Planck's constant.

Actually, according to James Clerk Maxwell, magnetism, light, and electricity are all manifested by the same phenomena, which is the electromagnetic field. He explains light as waves that constitute a combination of magnetic fields as well as oscillating electric. Einstein's

"Photon model" came into place when it was able to explain the photoelectric effect successfully. This effect has been explained in the next step.

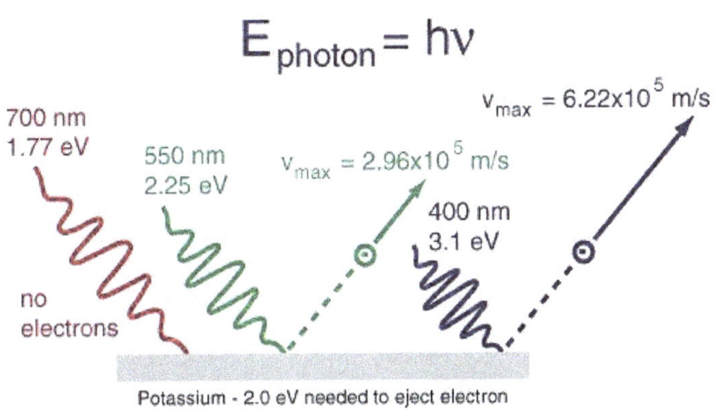

The Photoelectric Effect

According to Einstein's explanation, he argued that a beam of light had got the photons, which are particles stream and also a frequency "f"" The energy present in that photon will be equal to "hf"" This implies that there is no effect on the energy that relates to the beam's intensity. He further explained that to remove an electron from a given metal, "work function," which is a certain energy amount is required- it is denoted by "φ"" With his further explanation when the work function is higher than the photon's energy there will be no sufficient energy that is required to remove the electron from the given metal.

His description also argued that light is composed of particles that gave an extension of Planck's notion. This is the notion of energy that is quantized—whereby more or less amount of energy can be delivered by a given photon depending on its frequency. There was a compromise on the particle state of light since it was explained that light also had waves. This resulted in the consequences of the quantization of light.

Matter Quantization

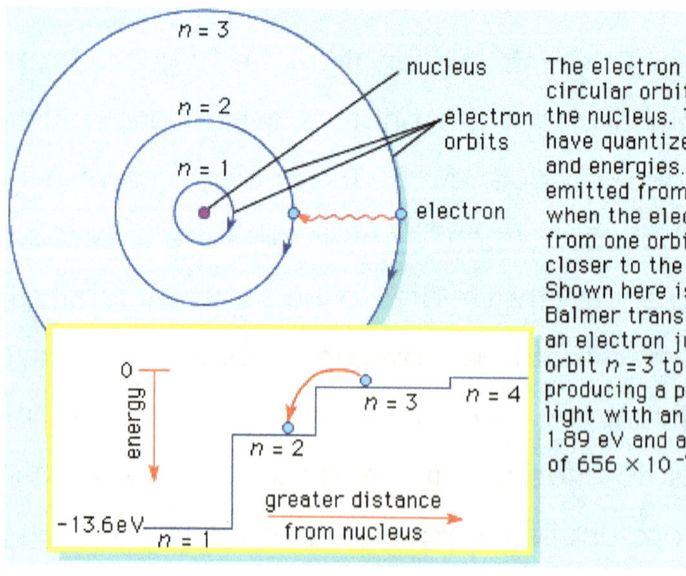

The electron travels in circular orbits around the nucleus. The orbits have quantized sizes and energies. Energy is emitted from the atom when the electron jumps from one orbit to another closer to the nucleus. Shown here is the first Balmer transition, in which an electron jumps from orbit $n = 3$ to orbit $n = 2$, producing a photon of red light with an energy of 1.89 eV and a wavelength of 656×10^{-9} m.

According to Bohr, the electrons jumped from one orbit to another. In this case, it gave off the light emitted in the form of a photon. The energies that are emitted by the photon were highly dependent on the differences in energy that were present between the orbits. In the first place, there were very many critiques in Bohr's model. Many argued that this model was wrong, although, in the end, it was evident that the model was good to suit quantum physics. With his explanation, Bohr argued that matter has also got some wave-like properties. According to him, an electron beam can also exhibit "diffraction." This is a similar case, just like the beam of light or a wave of water. Thus, small molecules and atoms have got the same phenomenon. To prove the above, a double-slit experiment was undertaken-this experiment is explained in the next step.

Chapter 2

The Birth of Quantum Physics, Theories

The history of quantum physics is a very important aspect of the new physics. Before we get started, it is essential to know that quantum physics history is also linked with the history of quantum chemistry.

They both began with several multiple scientific findings. For instance, the cathode rays discovered by Michael Faraday in 1838, the discovery of black-body radiation by Gustav Kirchhoff in 1859 and 1860, the recommendation affirming that energy states of a physical system can be discrete by Ludwig Boltzmann in 1877 and a host of others.

Quantum physics was coined in Germany by a team of physicists, including Wolfgang Pauli, Werner Heisenberg, and Max Born. It was done at the University of Gottingen at the beginning of 1920.

While everything started with Max Born's theory, others gradually picked up, and they were all applied to bonding, chemical structure, and reactivity.

Most importantly, Max Planck and Niels Bohr were the original founders of quantum physics. They were recognized when they bagged a Nobel Prize in Physics, specifically for their effort on quanta.

At the closure of the 19th century, scientists had formed added information about how light and matter act. However, it was nearly empirical because it was based on observations and not explanations on theory.

Nobody was aware of how spectral lines were created, and no human knows how light could pass through space as either a wave or a particle.

Subtle was the name for the nature of matter. Of course, nobody also understood why different elements have properties that show as periodic, and as well, no one understood the reasons for the formation of radiation and fluorescence.

Meanwhile, while we entered the 20th century, it was revealed that all the above assumptions, which were not able to be explained, were all linked to quantum physics, which was, at that time, a modern physical law that looked to be effective in the microscopic world.

As time went on, quantum physics revealed that even small objects act in a way that flouts and goes beyond common sense. In recent times, this has changed our knowledge of the world at large.

Even with its weirdness, the special predictions of quantum physics have never been a lie. Additionally, the theory was formally used to explain multiple principles, like the behavior of the sub-atomic particles.

Virtually every new technology has its hope on the laws of quantum physics to be effective, and this also includes lasers, transistors, and microchips.

Quantum physics has evolved since the time of Max Born, Bohr, Schrodinger, and others to what it really is today.

Quantum physics is an area of physics that deals with the study of small but important bits of matter. In other words, it deals with things on a very tiny scale. One example would be the behavior of atoms and sub-atomic particles.

In the early 1900s, many people were arguing about whether light could be explained using particles or waves.

This is because one theory says that light can behave like waves (a series of peaks and troughs), while another theory says that it behaves like a beam or stream of particles (a series of lumps).

These ideas came from the work of Isaac Newton and James Clerk Maxwell. However, in the end, neither theory could fully explain things. Then, in 1905, Albert Einstein came up with a new way of thinking about light that did not involve waves or particles at all.

To understand this better, let's take a look at what is meant by quantum mechanics (that is what Einstein was working on). These are small amounts of energy that are contained within very tiny particles such as electrons, protons, and neutrons. Each different kind of particle has its own set of specific quantum properties. Scientists have found that these properties can be explained if they exist in a state known as superposition.

In other words, the particle exists in two different states at any time. There is a 50% chance that it will behave like one particular thing, while there is also a 50% chance that it will behave like another.

An important part of this research is the way that these quantum properties can be measured or calculated. It means that we can calculate what the energy levels of the different types of particles are going to be, and then their behavior. We do this by using wave functions. A simple example of this would be the hydrogen atom.

The Schrödinger equation makes use of wave functions to predict where electron clouds are most likely to be found around an atomic nucleus. It does this by using the wave function of an electron to find its probability density. The Schrödinger equation works only if you use it in conjunction with Heisenberg's uncertainty principle, which states that it is impossible to get a measurement or calculate the exact location and speed of a subatomic particle at the same time.

Quantum mechanics also has some very strange effects. They include behavior that seems to be random and unpredictable. It can be quite difficult for people to accept that things happen without any real reason, but nobody knows why they occur, although scientists have discovered ways of predicting when this sort of thing is more likely to happen than not.

One of these strange effects is known as the observer effect. Let's say that we are observing something like a radioactive atom, and we want to know what is going to happen over the next hour. When we look at this atom, we are really changing it. This is because the act of looking at it changes its properties so that it behaves differently from when we are not looking at it.

Another one of the strange effects in quantum mechanics is known as quantum entanglement. This means that two or more particles can become connected in a way that forces them to have a relationship with each other, even if they are very far away from each other physically.

Quantum mechanics have also been used to explain how things like telepathy and mind reading are possible. It would be impossible for us to understand these things if it were not for quantum physics.

Chapter 3
Quantum Mechanics, Practice

Although it is more than one hundred years old, quantum physics is still a relatively new and young science.

As such, proponents of classical physics might simply not agree with many of the theories postulated by quantum physics (and if you look closely at all the arguments currently going back and forth, you might also understand why there is a healthy dose of skepticism involved in this entire affair).

Unless we are conditioned by quantum entanglement to do certain things because it is "written" (in the stars?), the healthiest thing you can do is try to understand both sides of an argument.

In the world of physics, the main battle is fought between the supporters of Albert Einstein and Erwin Schrödinger, who found some of the theories of quantum physics to be not just unrealistic and improbable but downright ridiculous at times.

If you want to find out more about the battle between classical physics and quantum mechanics, why we still accept the theories of quantum physics (and why they are still being studied in the most reputable research centers of the world), and what the future of quantum physics might be, then keep reading.

We do think, however, that it is healthy for you to at least read a bit about the main points behind this debate - so we encourage you to do it even if you are not particularly eager to, and the perspectives shown by quantum mechanics are far more exciting for you (which is completely understandable).

Einstein's Explanation vs. Quantum Mechanics

It is a pretty well-known fact that Albert Einstein was not a big aficionado of the quantum mechanics theories that were shaping up during his lifetime.

Time proved him wrong in some ways because some of the quantum theories are being proven step by step.

Beyond that, however, the questions posed by Einstein are still valid - and they provide quantum researchers with a point of orientation when it comes to the answers they are yet to give.

If Albert Einstein were alive today, he would probably have "converted" to quantum physics—because even throughout his life, his views on this theory changed.

More specifically, in 1935, his experiments revealed what he called "spooky action at a distance" - or, in other terms, quantum entanglement. He then continued his experiments, furthering the theory that quantum entanglement was only possible in certain circumstances. Unfortunately, however, he never got a clear answer to this follow-up, and it was left to future generations to reconcile the theories. It would be more than interesting to see what he would have to say about the more recent discoveries and experiments in quantum mechanics.

Why Do We Accept Quantum Mechanics?

There is no doubt that Einstein's works have reshaped the entire world in many ways so that the entire library of books only uses simple English to explain them. In the scientific community (and, dare we say, outside of it too), Einstein is seen as a sort of demi-god—an irrefutable authority that nobody dares to touch.

Nobody except quantum physicists, that is.

If Einstein's Theory of Relativity is so well-regarded and accepted, why do we even bother with quantum mechanics, then? What demon sets so many contemporary scientific researchers on the path of actually trying to reconcile the worlds of classical physics and quantum physics?

Well, the one reason quantum mechanics is accepted and still very much a topic of discussion is that it would manage to solve what classical physics couldn't.

September 7, 2014, might have seemed like any other day of fall in the Northern Hemisphere. The leaves were probably slightly yellow by then, and the heat of the summer was slowly starting to wear out. Maybe it even rained a little in the morning, and by the time cities were waking up to life, the fog of a slightly chillier night vanished, leaving room for a perfect day of autumn.

Everyone must know that September 7, 2014, was the day when the "Theory of Everything" officially appeared. You may have heard of it because there is a movie about the life of Stephen Hawking. Or you might have even stumbled upon it long before the movie came out.

What is important, however, is that the Theory of Everything is one of the most important attempts at unifying both the theory of relativity and the quantum theory. That which started in the 1920s by Albert Einstein was finally beginning to make sense eight decades later by the hand of Stephen Hawking.

The Theory of Everything is, perhaps, one of the most ambitious projects ever. It is one of the theories that is bound to change every single little thing - not just in physics, but in science as a whole, and, soon enough, in mankind's perception of pretty much every area of their lives.

What the Theory of Everything tries to do is finally build a bridge between quantum mechanics and the theory of relativity. Some would even dare to say that it will "tell the mind of God" (Marshall, 2010) and that it will hold the key to mankind answering the questions it has been trying to answer for a very, very long time now.

There are several candidates for the "Theory of Everything." Some of them are implausible to be actually proven in an equation or practice, but some of them stand out as sane options that might be the final answer to everything.

Out of these, we would like to take the time to name the two most important contenders.

One of them is called "String Theory," and what it says is that there is a ten-dimensional space we are living in. That sounds more than mind-boggling, we know, but wait until you hear more of it.

According to the String Theory, the point-like particles of particle physics are one-dimensional objects (called "strings"). The theory describes that these strings propagate through space and that they interact with each other. When looked at from afar, a string looks like any other ordinary particle (with a mass, charge, etc., that is determined by the vibrational state of the string). For instance, one of the vibrational states of the string is represented by gravitation (a particle that carries the gravitational force, that is).

In essence, the Theory of Everything relies on quantum gravity, and it aims to address a wide range of questions in fundamental physics - such as what is going on with black holes, how the universe was formed, how to improve nuclear physics, and how to handle condensed matter physics better.

Ideally, string theory will unify gravity and particle physics (which is one of the main points that have to be bridged between classical physics and quantum mechanics). However, it is not clear how many theories can be adapted to the real world and how many of them can change their details.

The other theory competing with string theory for the title of "The Theory of Everything" is the Loop Quantum Gravity Theory. This paradigm is heavily based on Einstein's work, and it was elaborated towards the middle of the 1980s. To understand it, you need to remember the fact that, according to Einstein, gravity is not a force per se but a property of space-time.

Up until the Loop Quantum Gravity Theory, there have been several attempts to prove that gravity can be treated as a quantum force, like electromagnetism or nuclear force, for example. However, these attempts have failed.

What the Loop Quantum Gravity Theory tries to do is to base the bridge between traditional physics and quantum physics on Einstein's geometric formulation. Ideally, this will prove that space and time are quantized the same way as energy and momentum are in quantum mechanics.

If physicists manage to prove the Loop Quantum Gravity Theory, space-time will be pictured with space and time being granular, which would consequently mean that a minimum space exists.

Although String Theory seems to be a lot more popular in mainstream media (mainly because some of its proponents are quite popular themselves, even well outside of scientific circles—like Michio Kaku, for example), the Loop Quantum Gravity Theory should not be dismissed in any way. Most of its implications are related to the birth of the universe, which is why it is also called the Big Bang Theory—and, perhaps, why the eponymous TV show was so named as well.

Chapter 4

Fundamentals of Quantum Physics and Laws

In today's world, we must possess a sound understanding of quantum physics and its implications.

What is the fundamental nature of physical reality? Why do particles behave the way they do in quantum mechanics? What are the laws of quantum physics? How does quantum physics relate to our subjective perception of reality? There are many fundamental questions about life and how it works that are not yet answered with certainty. Quantum mechanics may provide some answers to these questions.

For example, when you decide to carry out an action, such as pushing a heavy box across the room, your decision has an instant effect on what actually happens in our world. Of course, a very large number of neurons in your brain must carry out many complicated processes so that you will push the box. However, we can say that it is ultimately because you decided to push the box across the room that the box moves. On a fundamental level, we can say that the following sequence of events occurs:

1. You decide to carry out an action (push a heavy box across the room), and this decision has an immediate effect on what happens in our world;

2. A number of neurons in your brain carry out many complicated processes;

3. Eventually, your right arm pushes against the box, and it moves across the room;

4. When this happens, it is as if the box moves because of the decision you made long ago.

On the surface, it would seem that events 1 and 2 are separated by a time gap equal to the time required for your neurons to carry out their complicated processes. However, we know from quantum mechanics that in our world, there is no gap between events 1 and 2. Quantum mechanics tells us that events 1 and 2 are connected intimately and fundamentally. On a fundamental level, it seems that events 1 and 2 are connected by instantaneous action at a distance. This fundamental connection between your decision to push the box and how the box moves is a type of nonlocal connection.

To make sense of this type of connection, we will need to delve into the nature of reality on a fundamental level. Our ultimate goal will be to determine how events in our world are connected in an intimate and fundamental way.

Viewing Our World as a Hologram

One way to think about how events can be connected by instantaneous action at a distance is to view our world as being like a hologram. For holograms to work, they must possess a degree of spatial compression (to allow all the information about an object or scene to fit on a flat photograph). Besides, there must be a degree of temporal compression (to allow all the information about an object or scene to fit on a single piece of film).

The reason for viewing our world as being like a hologram is to make sense out of how events in our world are connected in an intimate and fundamental way. In this case, holograms can give us insight into how instant communication is possible.

If we were able to view the world around us as a 3-dimensional hologram, it seems that we could instantly "communicate" with objects in this world by simply altering the orientation of the viewing screen.

Now, we know from quantum mechanics that there is no fundamental gap between events in our world. In fact, we know that it is possible to measure a state of a quantum system and instantly alter the state of another quantum system, even though there is no time lag between the two measurements (in between the two measurements, both systems continue to evolve according to the laws of quantum mechanics).

Thus, it seems that there must be some structure in our world that allows instant communication between objects in our world. We have found this structure. Holograms give us insight into how instant communication is possible and how events in our world are connected on a fundamental level.

It is important to realize that the holographic model does not say anything about how the information in a hologram is encoded. Perhaps there is some fundamental law that dictates how information in a hologram should be encoded, and perhaps this law can be written as an equation. In any case, what we will learn is that the holographic model gives us insight into how events in our world are connected by instantaneous action at a distance.

To say it another way, we will first describe how everything in our world can be viewed as being like a hologram and then describe some of its implications. Then we will discuss some of the many different viewpoints that exist regarding quantum physics and consciousness.

Holographic Description of Reality

First, let us view events in our world as being like a holographic image. Just like a flat photograph can contain all the information about what is in front of the camera, and just like a single piece of film can contain all the information about what is in front of the camera, so it seems that our world can be viewed as containing all the information about events in our world (as well as containing the fundamental laws which govern these events).

The idea that we are just "particles" moving around has been shattered by quantum physics. However, we do have an illusion of individuality. We think we are separate individuals who possess free will and who make decisions for ourselves.

For example, when you push the box, you will have the illusion that you decided to push or not to push. However, this decision will be predetermined by events in your past. This is because our past contains the information that is required for us to make decisions in our present and future.

For example, suppose you have a habit of always pushing a box whenever it is nearby. Perhaps it was left there by an absent-minded friend, or perhaps it has been placed there by another person who likes to play jokes on others. Either way, you will find yourself pushing the box even though you feel like it was your decision whether or not to push it.

This is not to say that you are a puppet whose every move is predetermined by some hidden hand. You are free to do what you like. However, the only freedom you have is to make decisions that are determined by your past.

Thus, we can view the universe as a gigantic holographic movie that contains information about every event that takes place in our world and how these events are connected on a fundamental level. This movie is playing in perfect sync with itself, and we can watch this movie at any moment by simply altering the orientation of our viewing screen. All we need to do is understand how this screen works and how it can show us the parts of the world that interest us most.

The Holographic Screen

It is a fundamental property of all matter that it can be viewed in two ways. One way is to view it as an indivisible entity. This corresponds to viewing the world as though it were a movie where we see what is happening on the screen at one moment in time and cannot see what was happening in the past or will happen in the future. This point of view has been popularized by Descartes's idea of an immaterial soul which inhabits the body and sees only one moment, just as we view a movie on our television set.

The second way we can view matter is as a series of events that took place in our past, present, and future. This corresponds to viewing the world as though it were a holographic movie. A holograph is composed of two parts. One part is the object itself and the second part is a piece of film which represents the interference pattern between light waves that traveled from the object to your eyes.

We view our reality through a similar process. Light waves from objects travel through space and interact with our consciousness to form an image. However, instead of forming an image inside us, this image corresponds to a region in space where an event has occurred (a point in time). Light is arriving from the past almost instantaneously in all directions. When a light wave arrives at an event (a point on the screen), it interferes with other light waves which have traveled from different points in space to that event. These waves combine to form a holographic pattern which is stored in a thin region of spacetime. This process occurs continuously, at every location at which an event occurs and indeed throughout the entire universe, storing a record of everything that has ever happened since the beginning of time.

The holographic principle explains how our world can appear to be continuous and smoothly flowing through spacetime when it is really made up of individual events. The famous double-slit experiment illustrates this effect very well. When waves (such as water waves) travel through two slits, they form a holographic interference pattern on a screen. Similarly, when an electron goes through the double slit, it is diffracted in a way that can only be explained if the electron behaved like a wave and traveled through both slits at once.

The holographic principle implies that all of space is like the two-slit screen—there is no difference between our world and the subatomic world at all, except for our inability to see how each individual particle directly affects its environment. This theory can help explain some previously unexplainable events such as telepathy or why when we send one particle into another dimension, suddenly two particles appear elsewhere. According to the holographic model, this is not magic; it is simply a different way of looking at how our universe works.

Chapter 5

Conflict Between General Relativity and Quantum Physics

For the changing society of the early twentieth century, the fact that using only incredible mathematics aspects of nature could be described was a pleasant surprise. Perhaps not that it was an unprecedented experience, but few expected such a revolutionary change in a paradigm that had lasted for centuries. Before Einstein, the Newtonian description of gravity was revered for its simplicity and universal validity.

Relativity was soon discovered to hold more surprises. In 1915, the same year of its presentation in public, the theory revealed the possible existence of astrophysics bodies. Physics with stellar masses that do not let even light escape or black holes. These matter-eaters beasts immediately became the most seductive and enigmatic celebrities of relativity for a wide audience. Soon after, Einstein showed that his theory predicted the existence of gravitational waves, deformations of the cosmic tissue that move at the speed of light as if they were very fast space earthquakes. As if this were not enough, it was soon realized that a detailed scientific description of the history of the universe was possible based on the equations of relativity.

Despite the interest of these last entries, for several decades of the last century, the study of black holes, gravitational waves, and cosmology were seen as nothing more than curious theoretical works in more conservative scientific circles. The reason was and has been that it is difficult or impossible to make direct confirmations about these theories, and that it is difficult to make precise relativistic calculations to be able to compare them with indirect observations.

However, ingenious (and sometimes just lucky) experiments have accumulated vast indirect evidence. For example, in cosmology, the discovery of cosmic background radiation, predicted and described by the Big Bang cosmological model as remnant radiation from an early time when the universe was very hot, was an important building block in the consolidation of cosmology. This discovery was award-winning. Despite this, there were still doubts about whether the big bang model, based on general relativity, was correct to describe this radiation. Fortunately, a decade later, it was confirmed that radiation is not uniform throughout the cosmos and that the measurement of these inhomogeneities matches what the theory had predicted. Celebrated with the Nobel Prize in 2006, this discovery did not leave many options open. The cosmology described by relativity is the most appropriate description. A more detailed measurement of the expansion of the universe soon led to the last greatest cosmological discovery: the universe grows faster and faster (Nobel Prize 2011). And (almost) everything can be perfectly adjusted from general relativity.

The confirmation of the existence of black holes and gravitational waves does not have a shorter history. For one thing, black holes have always represented a singular scientific nuisance. The fact that the theory indicates that the gravitational force at the heart of black holes is infinite indicates a serious problem: right there, the theory of general relativity is no longer valid. Thus, for some time, it was considered that black holes were a trick that mathematicians played on us. However, theorists argued that, like stars made entirely of neutrons, black holes were corpses of stars larger than ours. The confirmation of the existence of neutron stars in the 1960s and indirect observations of the movement of stars around dark regions led to the certainty that there are many black holes in the universe and that they can have masses millions of times that of the Sun.

Additionally, black holes, while not the typical cosmic vacuum cleaners that we paint, do absorb large amounts of matter from their cosmic neighborhood, creating around them a ring of incandescent radiation-emitting material called an accretion disk. This radiation predicted by the theory has been confirmed, especially in the center of galaxies like ours. In the Milky Way, the movement of a group of stars around the center of the galaxy, "chased" by astronomers since the 1990s, has revealed that their lives a relatively small and dark body with a mass of almost 3 million times that of the Sun and that it emits radiation according to predictions for a hole. As if this were not enough, in 2011, an accretion disk was observed with the emission of X radiation, consistent with the predictions of black holes with masses of billions of times that of our star, absorbing material from a quasar.

The last series of indirect observations of black holes have to do with the gravitational waves. Although these can occur with any accelerated movement (even a clap produces them), according to the theory, only gravitational waves generated by violent cosmic events are capable of producing gravitational waves that we can detect on Earth, with sensors so sensitive as to measure deformations of the space of a thousandth the size of a proton or less. Gravitational-wave detection led to the award of the 2017 Nobel Prize. Supercomputers with sophisticated programs managed to show that the signals, according to relativity, were only consistent with the collision and mixing of two black holes with masses equivalent to a few dozen times that of the Sun. All at once, two of the most controversial predictions of relativity were proven just a couple of years ago! And the evidence continues to accumulate today to explore possible deviations from the predictions of Einstein's theory and find astronomical applications of the study of gravitational waves.

Not only gravitational waves and black holes are under the scrutiny of research. As we said, the cosmological model of the big bang can explain all observations of the dynamics of the universe under very simple assumptions on the geometry of space-time and an assumption on the content of the cosmos based on recent measurements, which is: 5% matter like that of our planet, 27% is a type of matter called dark matter that does not emit light, and 68% of cosmic content is a form of energy nicknamed dark energy that causes the accelerated expansion that we observe.

The biggest mystery is that no one has the slightest idea of what are dark matter and dark energy. It is nothing that we have been able so far to observe directly, although there is sufficient indirect evidence to affirm that there are such "substances" or something that has the same effects. There are those, however, who are convinced that we must slightly modify the basic equations of general relativity to understand the true nature of these dark entities. Others consider that only the immensely challenging search for the compatibility of general relativity and quantum mechanics will help us dissipate our doubts.

And this mix brings us today back to black holes. The inside of the black holes, being not observable, is totally unknown. All we know is that gravity must be so strong inside that it could have effects on the smallest particles, comparable to the effects of quantum forces which, according to particle physics, govern their behavior. If so, it is possible that a form of quantum gravity is manifested there, which we must theorize based on what we have verified in the last century.

But even for those least interested in the fundamentals of gravity, the theory of relativity today offers important modern tools. In addition to being relevant to the global positioning system (GPS), it is crucial in astronomical observations. The deflection of light due to its passage near galactic formations, stars, planets, etc., causes an apparent shift in the position of the stars and galaxies concerning the real one. But it is not the only effect. If there is a galaxy behind a very massive astrophysical body, the deflection of light beams emitted by the galaxy in all directions can be deflected towards us around the contour of the galaxy.

Astrophysical "nuisance." This effect is called gravitational lensing. Gravitational lenses not only allow us to characterize what is behind the observable objects that cause the deflection of light but also when they occur in regions where there are no visible obstacles, they show us properties of unobservable objects, such as black holes and dark matter formations, which have not yet been fully described.

Relativity, despite its age, remains a developing treasure, whose questions and responses pose current challenges that are likely to become the basis for future discoveries and a paradigm shift like the one Einstein witnessed.

Quantum Mechanics and General Relativity Incompatibility

Physicists are trying to reconcile the two textbooks according to which science understands the world - but to date, there has surfaced no proven, palpable theory to bring the two worlds together and finally help us understand where we come from, where we are, and where we are going (because, at the end of the day, these are the fundamental questions both classical and quantum physics propose).

In classical physics, according to Einstein's principle of general relativity, the reality is composed of 4 dimensions, called the space-time continuum. Gravitational fields are continuous entities, according to this paradigm.

In quantum mechanics, however, fields are not continuous but discontinuous. They are not defined by the 4 dimensions, but by "quanta" As such, concepts like the "gravitational field" are missing from the world of quantum physics, which is also the biggest bridge classical physicists and quantum researchers have to build between their points of view.

This is not just a matter of fancy definitions. The world of quantum mechanics and the world of classical physics are incompatible because they describe reality in completely different ways, in dissimilar terms, and unlike perspectives that do not meet at any point.

In classical physics, all events that happen, occur for a reason. They come about according to the old dictum of cause and effect. Nothing happens randomly, but because something has caused it to happen in the first place.

In quantum physics, the reality is not observed in terms of cause and effect but in terms of particles jumping from one state to another based on probability, not on definite outcomes.

It is, therefore, worth thinking about the importance of achieving reconciliation between classical physics and quantum physics. Even when these two disciplines seem so different and at such a deep level?

This is because reconciliation would allow complementarity between the two disciplines. Where classical physics fails to provide explanations of the microcosm, quantum physics would succeed. And where quantum physics fails to provide an answer when it comes to the macro-objects, classical physics would bring logic to them.

Chapter 6

Quantum Physics, Development and Technology

Theories of Quantum Mechanics have been around since the early 20th century, but it has only been in recent years that they have been put to use. The reason for this is that early quantum theories were mainly mathematical tools for calculating and predicting the correct behavior of a system. The thing is, Quantum reasoning concerns something called a quantum state, which itself gives off its light and hence cannot be observed without changing the original state, so this means we are never really going to know how things would work in a 100% pure sense. This makes it difficult for scientists as they can study particles and light and from this deduce how quantum interactions take place but cannot always prove that this is how they actually happen; there are, of course, theories on how these works, but as of yet nothing is 100% certain. The main reason for this is that quantum states are very different from the normal world and have an element of uncertainty about them, but something else to be taken into account is that we humans just aren't smart enough to fully understand it all.

However, there is a great deal we can do with imperfect knowledge. One of the biggest applications for quantum theory is in the area of quantum computing, which has been making some progress over the last decade and is mostly focused on building a quantum computer that could solve an NP-complete problem in polynomial time. This essentially means that a quantum computer would be able to solve difficult problems very quickly. The reason why it is so hard to build a quantum computer is due to some of the strange things that happen when working on atomic and subatomic levels.

In order to try and make sense of this, I will have to explain how Quantum Physics works; if you find it too difficult then, please skip ahead as I know it gets kind of heavy at times. The first thing to know about is the Quantum State, which is itself a set of values for any one particular property of the particle; this could be the spin or energy level; it differs from regular atomic states as mentioned earlier, as they cannot be observed without changing them. This means that the

Quantum States is something that we can only know a probability for, and this, in turn, leads to two strange things; the wave-particle duality and entanglement.

The wave-particle duality is the idea that every particle has a chance of behaving either as a wave or as a particle, depending on how it is observed. In practice, this means that if you measure a particle, you will not get a solid value but a range of values all at once. In the same way, if you were to take a snapshot of an object with a camera, the picture that came out would be blurred as it is recording wave-like characteristics. This, in turn, leads to entanglement, which is what happens when two particles are observed in relation to each other. The idea here is that by measuring one particle, we affect the properties of another, sending it along quantum paths until they are observed again, and this makes them mutually dependent on each other. It is the sheer fact that they are entangled that makes it so difficult to understand quantum states or to predict how they will behave as they can only be affected as a result of their entanglement; this means we have no definite values for them, but instead we have a range of values in which the particle could be.

In my opinion, a lot more work needs to be done here on the application of Quantum Physics, but I do think it is a step in the right direction. It really does seem like it has opened up a new world of possibilities for our species, and I am excited to see where we go from here on out, especially with all this talk about intelligent robots.

Chapter 7

The Reality Seen From the Quantum Physics

At a point, we must feel irritated and slightly confused as to how so many concepts been discussed here can be applied to our daily lives or used by the different instruments around us. Quantum Physics is one of the highlights of human intellectualism, and its knowledge has really helped shape our civilization. Despite this relevance, most people still feel the subject of this field is quite abstruse and cannot be easily grasped by the ordinary mind. In the mind of the public, the concept of quantum physics is seen as a hard concept that is only understood by minds like Einstein and Hawking and another superhuman brain.

The concept of quantum physics is an understanding of the universe, and the universe is all around us, and its operation is based on quantum rules. Even though we are so used to the laws of classical physics, and this relates to the universe at a macroscopic level, the understanding of quantum physics still affects various familiar operations. You'll find this list contains various tools and equipment that apply to the quantum principle without us realizing it.

Toasters

We are all familiar with the red glow produced by the heating element as we toast our bread. Funny enough, it was the observation of this red light that led physicists to ask questions that gave rise to the quantum concept. Physicists wanted to know why hot objects shone that particular color of red, a very tough question, and quantum physics came to provide light to it.

The color of light which is emitted by a hot substance is a universal phenomenon that occurs irrespective of the substance that has been heated, and notwithstanding the extent of heat it can take, it will emit a similar spectrum of light as any other substance. This observation stirred the curiosity of many physicists in 1800 who made attempts to study it but failed to provide a reputable solution to the problem.

The major point that the light was independent of the composition proposed a simple universal approach: you count the number of colors of light that can be emitted by an object and then provide an equal portion of the heat energy contained in the object. It was also observed that high-frequency light than it was possible to emit low-frequency light; this meant that the roster could emit very x-rays and gamma-rays other than the warm red glow. This wasn't the case, and it meant another phenomenon was taking place.

Max Planck helped provide a solution to this problem in his hypothesis where he said that the light been emitted could only be emitted in discrete chunks of energy, integral multiples of short constant times the frequency of the light. For high-frequency light, the energy quantum is greater than the share of heat energy, which is assigned to that frequency, and this makes it impossible for light to be emitted at such frequency. This prevents the emission of high-frequency light. We could say that the toaster could be a major place where the idea of quantum physics first originated.

Fluorescent Lights

The traditional incandescent light bulb was able to emit light by heating a piece of wire adequately until it gets hot and emits a bright white glow. This is similar to the phenomenon of the toaster. You are enjoying a groundbreaking work of quantum physics whenever you flick on a fluorescent bulb or one of the more recent twisty CFL bulbs; that's quantum physics at work.

In the early 19th century, physicists discovered that all elements found in the periodic table have a very unique spectrum. When we heat a vapor of atoms, they will eventually emit light at a smallish number of discrete wavelengths, and each of the different items will have a different pattern. The spectral lines were used to classify the composition of unknown material, and unknown elements such as helium were first discovered through this process.

This is how a fluorescent bulb works: whether the bulb is CFL or a long tube, inside the bulb is a tiny bit of mercury vapor that is excited into plasma. Mercury easily emits light at frequencies that fall at the visible spectrum, and eyes will perceive this as white light.

Chapter 8

The Physics of Possibilities

Quantum physics can be a complicated subject. It's been said that quantum mechanics is the physics of possibilities. To reach quantum conclusions, you need to ask the "what if" questions and follow how a particle will behave. Quantum also has the concept of uncertainty, which is an odd thing considering that these are particles in some sense knowable. To enjoy the simplicity of quantum physics, let's start off with a simple example.

If you have a particle that has an even mass number and you want to find the location of that particle, you can perform two experiments. In experiment one, the particle doesn't move, and in experiment two, the particle moves towards its right. We can begin with these experiments and then make some assumptions about where the particles will end up based on probabilities or uncertainty.

The first experiment is to point a stream of subatomic particles (electrons) at a sheet of metal that has two slits in it, then counting how many particles go through each slit. In the second experiment, we will have slits, but this time with a magnet to move the particles to the right. The reason for this is that we know that the magnet will be attracted by a positive charge and pushed away by a negative one, so if both of our particles have a negative charge, we know that it will be pushed towards its own right.

The results of experiment one would lead us to believe that if you drop an even number of molecules (particles), then half of them will go through the hole in the left and half through the hole on the right. In stark contrast, in experiment two, it would seem that every single particle goes out only one hole.

As physicists, we can assume that half of the particles go through both holes and that the other half is blocked. We then make the assumption that every particle wants to go through one hole and one hole only. The fact that we count more than half of the particles going through a single hole is because some want to go out both holes, but there's no way for us to know exactly how many went out both holes, so it will contribute to uncertainty. It would be even more uncertain if we used larger or smaller particles because they are harder or easier respectively to detect (remember, we're talking about electrons). In smaller particles, it's easier to miss the mark.

Now that we have a better idea of the probable location of a particle, let's look at another example. Say you want to predict where an electron will appear on a screen when someone flashes a light at it. The experiment goes like this: you fire electrons at a screen that has two vertical slits in it (one for each slit you fired electrons out).

Because you fired the electron from one point, there is zero uncertainty in its starting position. You will see that each electron behaves differently and goes through different holes, but every single time they will go through one hole and never both holes.

After the experiment is over, you go back and look at your data. You notice that the electrons act in a strange way. Not only do they not behave like waves, but they appear to be able to know in advance where the second slit is.

Quantum mechanics look at where electrons will appear on a screen based on probability. We can assume that every electron wants to go through both slits, but it's impossible for them to be in two places at once, so they have to appear by chance. We can assume that the electrons aren't in a wave state, so they have to appear in a certain place out of all possible places.

Because there are two slits, we know that there is going to be interference, and it will look like the electrons are waves. We also know that because we know where each electron came from, there is zero uncertainty in its starting position. Also, because there was only one electron fired at a time, there is no uncertainty with regards to how many electrons went through which slit.

In this experiment, we can assume that each electron will go through one of the two holes. We then devise a mathematical equation to show this. In the quantum world, things are bound to happen because of probability, so the probability wave theory comes in. A probability wave is like an invisible force that tells the particle where and when it should appear on a screen.

This concept is called the Copenhagen Interpretation, but it has been revised over time as people have come up with new methods to test quantum mechanics. The first iteration of the Copenhagen interpretation came from Niels Bohr in

1927 and was named after his hometown of Copenhagen, Denmark (it's also known as the cradle of quantum physics).

The Copenhagen interpretation tells us that, at the very core of everything, there is a particle that has a nucleus made up of protons and neutrons, and it has electrons orbiting it (that might be tense due to gravity). These electrons are all over the place in a haze of uncertainty.

In order for an electron to be identified as an electron, the particle will have to go through one hole and not both. In order for this to happen, you need an observer or detector. You can imagine that if you had no detectors (e.g., eyes) then you would see a probability wave interference pattern on your screen with no definite measurement results.

Now, let's imagine that you have a detector. The observer, or the detector, has to look at the screen in order for a particle to be identified as an electron. By looking at the results of the experiment and comparing it with what data you already have, you can make an educated guess on where each electron came from and therefore predict where they might go.

That guess is based on probability because each electron acts like a wave with two different destinations (one for each slit), but electrons are particles, so they aren't allowed to go through both slits at once, and so they will act like a particle (a particle is something that never acts like a wave).

Feynman came up with a way to save electrons from going through both slits by looking at each electron as an event. By doing this, you can break down the wave function into smaller components that are all causally related (i.e., the electron will end up in one of these spots purely because it has been broken down into smaller pieces).

Since there is no such thing as time in quantum mechanics, there is no way for us to know if an electron has gone through a slit or which slit (and how could we?). All we can do is measure what its path is once it's on our screen.

One of the reasons why this makes sense is because what we see in our universe, at the macroscopic level, is mostly based on classical physics, which is the physics of big things like objects and people. If you try to apply quantum mechanics to these giant objects (like a hand that you place through a slit), then your results will make no sense and have no explanation.

The von Neumann (1956) formulation of quantum mechanics explained that all processes in the universe could be broken down into elementary units of atomic components. These atomic components are called atoms or elementary particles, and they're what make up everything we see around us. This is also known as wave-particle duality.

Chapter 9

"We Are All One": Interaction of Matter

In 1900, German physicist Max Planck demonstrated that radiant energy is made up of particle-like components, quanta, and particles can have wavelike characteristics. Quantum physics was born, and Max Planck is widely regarded as its father.

In this chapter, we will cover:

- Young's double-slit experiment

- Heisenberg's uncertainty principle

- The electron cloud model

- The heart of quantum mechanics

- Pauli's exclusion principle

- Quantum fields

One very important concept in quantum theory is Pauli's exclusion principle. Another is Heisenberg's uncertainty principle. But we'll begin by exploring the famous double-slit experiment.

Young's Double-Slit Experiment

In the 17th century, Isaac Newton concluded that light is carried by corpuscles (particles). Christiaan Huygens argued that Newton was wrong and light is a wave, but Newton's theory was accepted because of his greater prestige.

In 1801, long before quantum physics was even thought of, Thomas Young performed his famous double-slit experiment showing wave diffraction and interference, proving that light is a wave. The double-slit experiment requires a monochromatic light source, i.e., a single wavelength of light. It is unclear how Young achieved this as, unfortunately, he never recorded his process.

Different frequencies of the light spectrum make different interference patterns. All the frequencies together (white light) would produce a very blurry composite of all the patterns in this experiment.

Diffraction of waves can be observed when water flows through a gap or around an obstacle. You see diffraction as the water bends around the edges of the gap or obstacle, producing waves as it emerges. Light performs in the same way.

In Figure 1, when the light passes through the double slits in the barrier, we see two distinct waves emerging. When the two waves emerge, they interfere with each other in two different ways, as shown in Figure 2.

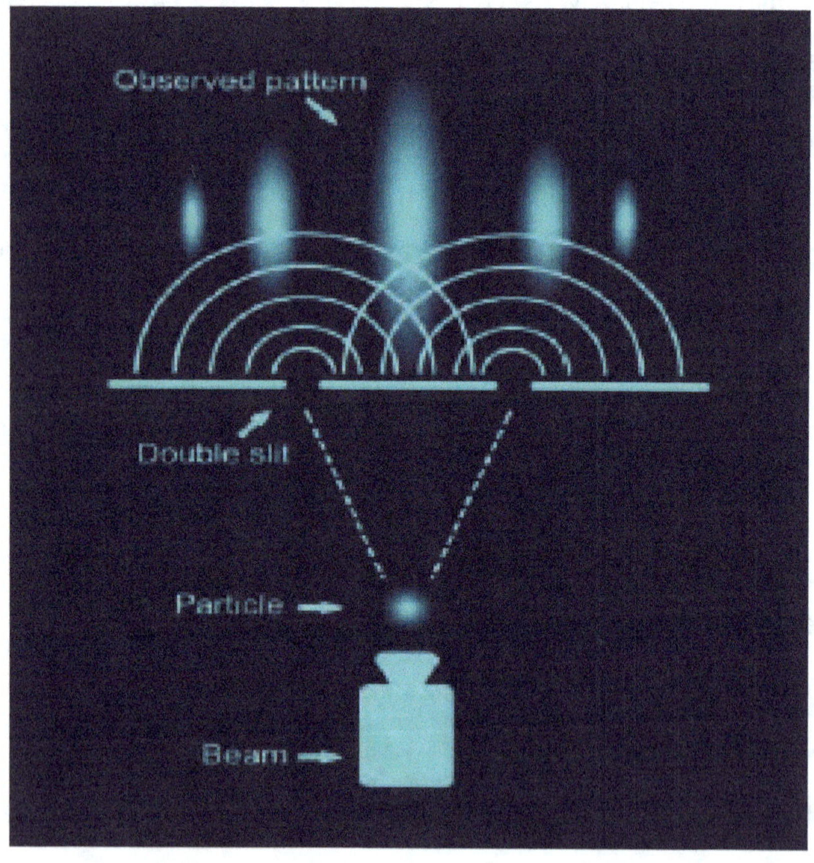

Figure 1/Photos by D/Shutterstock

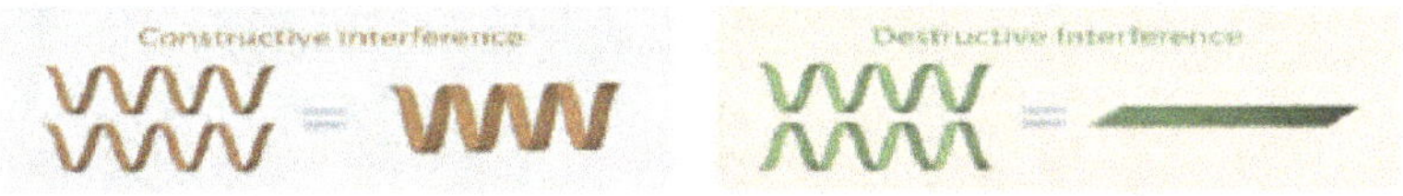

Figure 2/ Naski/Shutterstock

Where the crest of a wave meets with the trough of the other, they are reinforced. This is constructive interference. Where the waves meet crest to crest, they cancel each other out. This is destructive interference.

Where there is constructive interference in the light wave, the screen at the back shows bright stripes. When there is destructive interference, there are dark stripes on the screen (see Fig 1). This indicates that light is a wave.

You may conclude from this that Huygens should have won the argument with Newton in the 17th century. But read on.

In 1905, Albert Einstein, in his work on the photoelectric effect, proposed a quantum of light (the photon) that behaves like a particle and a wave. In other words, light is both particle and wave. This can be demonstrated using another version of the double-slit experiment (Fig 3).

Diffraction and Interference

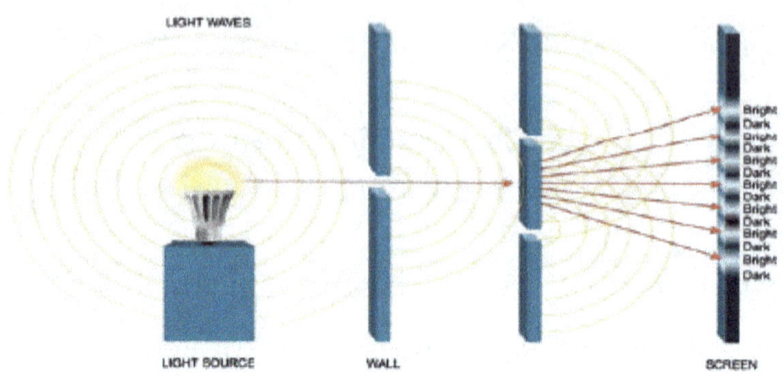

Figure 3/ Designua/Shutterstock

When monochromatic light passes through the single slit in the first barrier, it diffracts and emerges as a wave but appears to go through to the barrier behind in a straight line as a particle.

If we remove the first barrier and send light through both slits, the interference pattern of waves appears. If we now close one slit and send light through, it behaves as a particle as before. Einstein was right - light is both wave and particle.

In 1927, Clinton Davisson and Lester Germer showed experimentally that electrons perform like waves. Their experiment showed a diffraction pattern when electrons were scattered by the surface of a crystal of nickel metal.

This supported Louis de Broglie's hypothesis of 1924, in which he postulated that matter had wavelike properties.

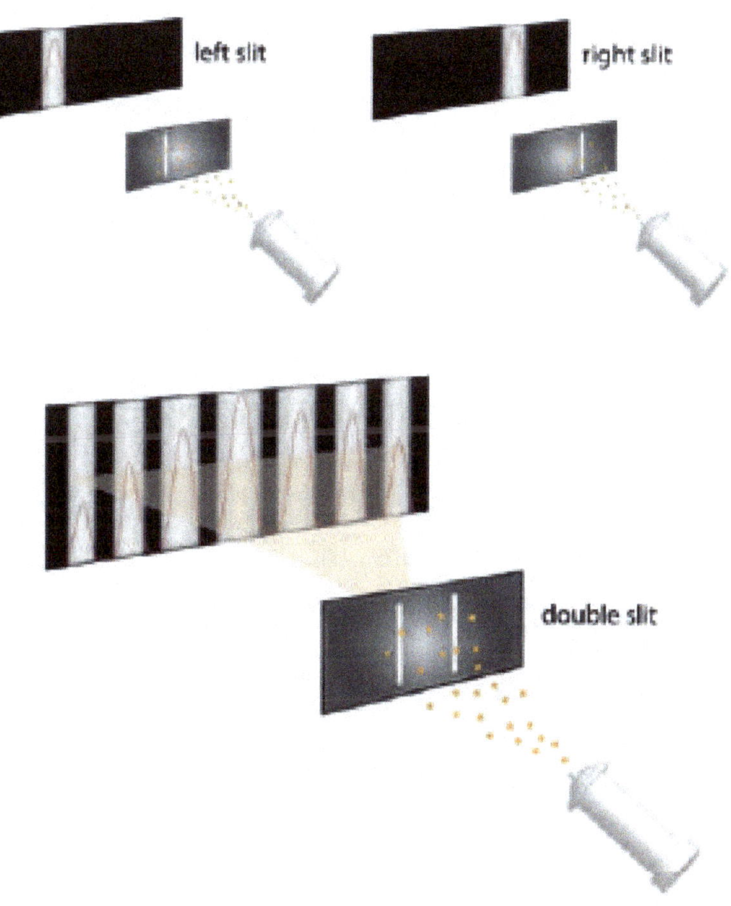

Figure 4/ Magnetix/Shutterstock

The double-slit experiment was first performed using a stream of electrons by Claus Jönsson in 1961. Diffraction and interference were displayed just as for light, indicating that electrons could be both particle and wave (see Fig 4). The double-slit experiment can now be performed using neutrons, atoms, and some molecules.

Antimatter was shown to behave in the same way as matter by physicists in 2018. They developed a double-slit experiment using positrons (antiparticles of electrons). The double-slit experiment can also be used to demonstrate how uncertainty and probability are prevalent in any quantum system, as we will soon discover.

Heisenberg's Uncertainty Principle

The uncertainty principle, introduced during the mid-1920s by Werner Heisenberg, states that:

Both the location and the momentum of a particle cannot be known simultaneously.

We can know the path of an electron as it moves through space, or we can know where it is at a given position. But we cannot know both. If we observe where electrons are, we cannot know their momentum and vice versa. We can only state probabilities of where particles may be or what their momentum is. This sets a limit to what we can predict in quantum physics.

The Electron Cloud Model

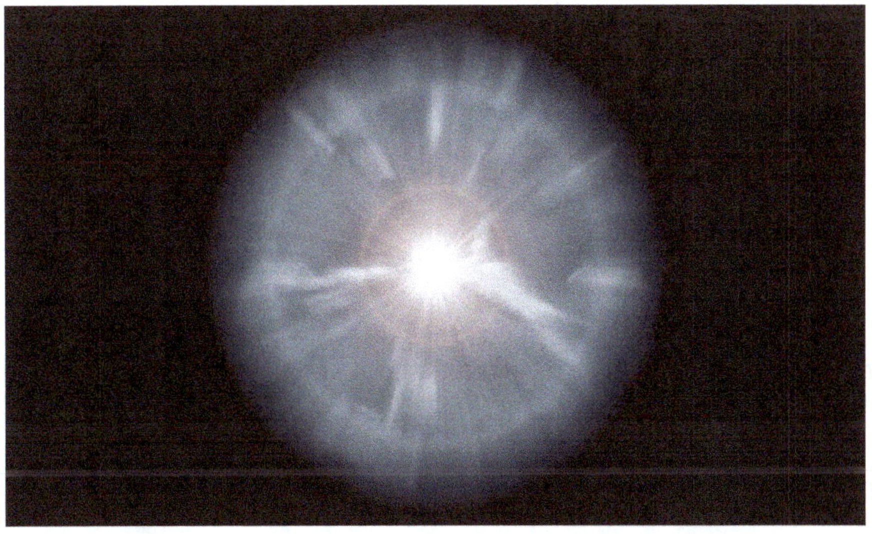

Figure 5 (artist's impression)/ Magnetix/Shutterstock

The electron cloud model was introduced by Erwin Schrödinger in the mid-1920s. This current model of the atom predicts clouds of the probability of where the electrons are around the nucleus. The exclusion principle is easier to visualize using Bohr's atomic model.

Let's take another look at the double-slit experiment to see how observing (measuring) an electron affects it.

The Heart of Quantum Mechanics

The double-slit experiment can now be performed to show the duality of matter and how observing the experiment can affect the result. The following was a thought experiment in 1978. It was performed using modern technology in 2007.

Figures 6 and 7 show the results of observing (measuring) an electron's position. *A device that fired electrons one at a time was used*—firing electrons one at a time with both slits open results in the interference pattern appearing on the back screen, just as it did with a stream of electrons. Every single electron appears to go through both slits at the same time as a stretched wave (see fig 6).

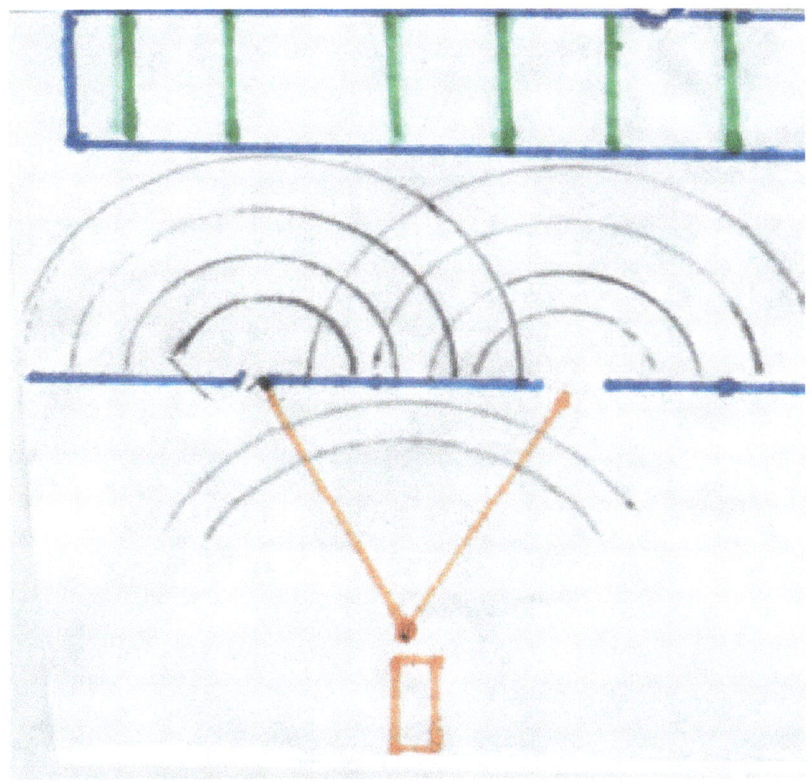

Figure 6/ Author's own image

If we now place sensors to observe each electron passing through the slits (Figure 7) we find that 50% of the time, the electrons go through the left slit and the other 50% of the time they go through the right slit. But the interference pattern does not show on the back screen. What we see are two stripes of electrons on the back screen directly behind each slit.

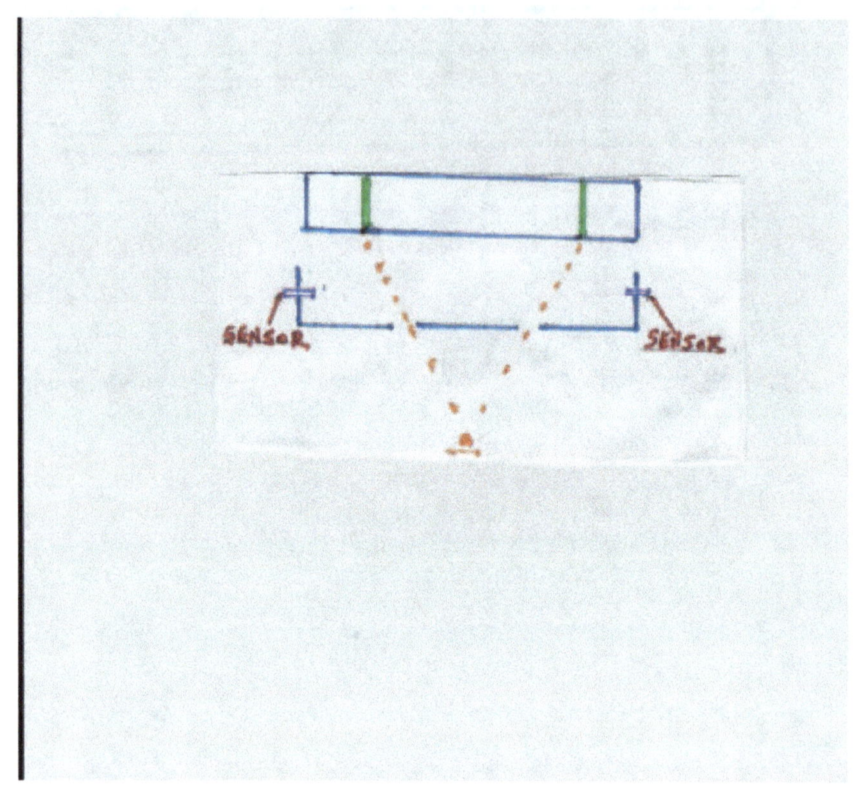

Figure 7/ Author's own image

The only difference in the experiment this time is the addition of sensors to observe the electrons. It's as if the electrons know they are being watched because if we remove the sensors, the interference pattern will appear on the back screen again. Are you confused? If you are, you're in good company.

Asking how that can happen demonstrates the measurement problem in quantum theory. Richard

Feynman, one of the most brilliant physicists of the 20th century, is quoted as saying: *"We choose to examine a phenomenon which is impossible, absolutely impossible, to explain in any classical way, and which has in it the heart of quantum mechanics."*

Pauli's Exclusion Principle

If atoms are almost all empty space, why can't we walk through walls? This question was answered by Wolfgang Pauli in 1925.

We now know that electrons, like all matters, can be particles or waves. The exclusion principle applies to all the fermions in the standard model but not to bosons.

Electrons have four quantum numbers:

- The principal quantum numbers
- The orbital angular momentum quantum number
- The magnetic quantum numbers
- The electron spin quantum number

The exclusion principle states that no two electrons in an atom can have the same four quantum numbers.

We learned from Bohr's model that electrons orbit the nucleus in fixed shells. And we learned that all elementary particles of the same type are identical to each other and in pristine condition. But their quantum numbers differ.

Atoms differ from each other not because they have different types of electrons but by the number of electrons they have and how they are arranged. An atom that makes up a solid will never allow another electron to join it.

Let's look at the atom of iron:

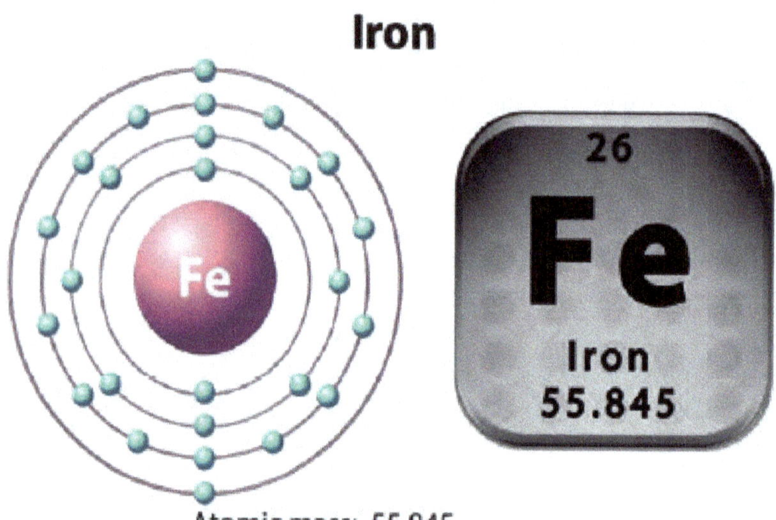

Atomic mass: 55.845
Electron configuration: 2, 8,14,2

Figure 8/ *Banderlog/Shutterstock*

Pauli's exclusion principle prevents unwanted electrons from entering an already made atom such as iron. But how?

If electrons were little round balls, it would be difficult to see how. But as we know that they can also be waves, then how they exclude unwanted electrons from an already made atom becomes easier to see.

Electronic waves can stretch to great distances but can never overlap. The outer (or valence) shell in the iron atom has two electrons. We can think of the stretched waves between electrons as barriers to keep unwanted electrons out by preventing them from overlapping. That is why the atoms in your feet could not force themselves through an iron floor you were standing on or any other solid.

Quantum Fields

We will not be looking too closely into quantum field theory (QFT) in this book, but I recommend more exploration in your future reading.

The bosons operate in force fields that permeate all "empty" space everywhere in the universe. Indeed, this is true of all fundamental particles. In this sense, there is no empty space anywhere in the universe, but everything in the universe is connected. We and everything are immersed in a sea of invisible fields.

Chapter 10

Human Consciousness and Thought Are Considered as Physical Entities

We just entered the age of the Aquarius. This means that our Solar System has moved into a completely different place in the Galaxy. We, the people living on planet Earth, have never been here before. What we can expect from this new space that we occupy remains to be seen. At the beginning of any great cycle, there are always harbingers, clues, and broad stroke rules for what we can anticipate. We've only been living in this place in space for fifty years. We've still got another 1950 years to go. What this new "Day" has revealed to us is already very important in the path that we should take.

First of all, the "Pisces Age" we've just left is over. Although many of the theories, structures, and forms of the Age of Pisces continue to be strong and alive, their plug has been pulled. They are no longer linked to the power of the Galactic Sun. The Galactic Sun is now driving the Aquarius Age.

The Age of the Pisces was known as the Age of "I Believe." It served its purpose because humanity needed to build its belief system from the 'Authorities.'

The Age of Aquarius is the Age of "I Know." We will be able to get all the knowledge we need to grow and evolve directly from the source. There's no need for a middleman.

This is made possible by one of the first gifts given to us by the "Day of Aquarius." It is the "Age of Mind," the gift of

Quantum Physics is for the mind. These laws will lay down guidelines for us to use for the next 1950 years.

They simply tell us that there is an infinite ocean of thought, intelligent energy called the Quantum Ocean. In fact, it's the mind of God.

All that has ever been, is or is going to exist there. There is no time, past, present, or future. Only the one NOW. There's no room here. No Width, no Length, no Depth, only the HERE.

Actually, the Quantum World, the Mind of God, is an infinite point called the HERE-NOW. And how we can seemingly live within this infinite point as well as outside this infinite point on Planet Earth is still a mystery to our finite minds. It will become glaring to us as we pass through the 2000 years. For now, we need to use the Laws of Quantum Physics and the concept of the Quantum Ocean, the Mind of God, to recreate a new reality for ourselves and mankind in general.

Another theory is that the Quantum Ocean, the Mind of Nature, is open to our "Thoughts." Thoughts are things! We need to really do more than just think about the Quantum Ocean, the Mind of God. We need to learn how to get there. How to be right there. We need to learn how to coexist 24/7, both psychologically in the Quantum Ocean and physically on the planet Earth. We need to understand how to use the Laws of Quantum Physics in the Quantum Ocean to restore our life and our world.

Quantum Physics and Your Health

The Laws of Quantum Physics posited that everything is energy. There is really only one basic force, and it resides in the Quantum Ocean. Scientists are banding back and forth theories and postulates about what is the basic building block of Creation. They're calling it "Quanta." They're talking about particles and waves, about possibilities and odds. Words, Words, Words! Definitions are less abstract than the object itself.

The term "apple" in a piece of paper is not a tree. You can't eat the word "steak" written on a piece of paper. The energy that occurs in the Quantum Ocean and 'blinks in and out of' our physical reality is the basic building block of Creation.

Where did this come from? God, or more precisely, the Quantum Ocean, is the Mind of God. It's the one thing that materialistic scientists leave out of their wit, talking and braying about Quantum Physics.

They will not come up with a correct answer until they place the Creator, God, in their equation. As to who we are, where we came from, and where we are going, the 'Big Bang Theory' concept does not have the correct answer to these questions. It doesn't answer the question, "Who put the energy in the ball and started the momentum in the first place?" The Creator, God, did.

When we know more about the Laws of Quantum Physics and the depth and breadth of the Quantum Universe, the true reality of the Creator, God, will be too immense for us to comprehend with our finite minds. We need to explore the Quantum Ocean, the Mind of God, and work individually with the energies of the Creator.

All of the Laws of Quantum Physics point to the intelligence behind Creation. The Quantum Ocean is known to be an infinite Ocean of Thought, Intelligence, Energy that responds to our thoughts and emotions. It's the intellect behind the development.

Since there is a Creator, therefore, there is also a design of the Maker. This concept is in the "mind" of the Creator. And because the Mind of the Creator is the Quantum Ocean, there is a strategy inside the Quantum Ocean.

Everything that has ever been, is, or will ever be, is contained in the mind of the Creator, God, the Quantum Ocean. It is in the form of the Divine Blueprints, the Divine Archetypes. These patterns are set up in such a way that when they 'blink out' into the Loom of our physical reality, they take definite forms.

There are Divine Blueprints for the Perfect Animals, Men, Women, Birds, Planets, Galaxies, and the Universe. When these Divine Plans initially 'blinked out' of God's Mind, they could be conceived as the Big Bang. It was fast. Each of us is an individual soul within the mind of God. We're 'blinking out' and 'in,' Birth, Death, just as the universes are. There's a Divine timing scheme for all of this. This is what the Bible's reference to "There's a Season" means for everything!

To achieve better health, we must aspire to balance our present energy configuration (mind and body) with the original energy configuration of our Soul. We should learn to enter the Quantum Ocean, search for and conform to our individual Divine Blueprint. We've got to burn off all the wrong "Dross" we picked up on our journeys. The Quantum Ocean is the place where the blueprints of perfect health exist. Time to enter and retrieve the Quantum Ocean, the Mind of God.

Quantum Physics and Its Connection to Your Self-Esteem

No matter what your beliefs are, the universe has secrets that cannot be explained and behaves in a way that is far from haphazard. It was Albert Einstein who said, "All this does not happen at random," and science is now moving more and more towards the idea that the universe is behaving intelligently and that there is a source of energy that shapes it.

The latest understanding is that the universe is made up of Dark Matter, Dark Energy (they don't know what it is), and Subatomic Particles in the form of matter and stuff that fills the void or gaps between matter, i.e. Space. We're the Energy!

Sub-atomic particles are not particles as such; they are particles of energy which, when observed, turn into solid objects, that is the matter we see in our universe and our everyday life. Albert Einstein's formulae, $e = mc2$ tells us that e (energy) = m (mass) x c (light speed) is squared.

This tells us that All Mass is equal to Energy. ALL MASS is equal to energy, not just sun or gas or oil or uranium. ALL MASS = Energy. It shows that everything in the universe, galaxies, stars, planets, moons, gases, and the US is energy. We're not made up of "we are stuff" energy (there's enough energy in the human body to fuel a small town for two weeks).

Such sub-atomic particles do not just sit there in mass to form matter; they vibrate, float, and morph in and out of the body in waves, forming and re-creating in a dance that is directly linked to changes in observation. They have intellect and fly through time and space without hindrance. They exist in the past, present, and future, and at present, they will respond and build based on information gathered from all these dimensions.

They don't judge what's right or wrong, and they'll form whatever you ask for, but they'll give you warnings and lessons to guide you along your path if that's what you need. But, of course, it's up to you whether or not you accept this and change your desire to reflect on the messages you're receiving. It's called Free Will!

It is said that everything that can exist already does and that it is through our observation of it (through thought) that creates it in our experience.

Take Thomas Edison, who performed thousands of experiments to discover the light globe, to give you an example. Using the Law of Attraction and by focusing on the end result he wanted, Edison knew all he had to do was persevere, and he would eventually find the right formula to invent the light globe. The light globe always existed as a possibility but only became a reality when Edison finally observed it as it should have been after thousands of experiments.

By the laws of attraction (The Secret), he knew that he would eventually achieve his goal. Why it took thousands of experiments to make this happen would have a lot to do with the thought processes that went through his mind at the time. It could have been his ego, his doubts, or his limited beliefs, but one thing is certain that, through persistence, he eventually achieved his goal, and the light globe became a reality as soon as he finally saw the globe as it is. That is when he was finally vibrating at the same frequency as his goal.

When we concentrate on thinking, experiencing, or believing in what we want from the sub-atomic particles that make up everything in the world, including space, we begin to work together and build it in our lives.

What we want is always a possibility, and the role of the world is to give it to us and turn back to satisfy us for whatever we want. It wants to give you what you want; as a matter of fact, that's its purpose. Forces far beyond our understanding come into play, and the arrangement of events begins to move into a position that will lead to circumstances that will give us all we want.

Chapter 11

The Power of the Mind

The quantum world is a strange and complicated place. Everything in the universe can be broken into its smallest parts, from the quarks that make up protons and neutrons to the constituents of light and even time itself. The first thing you need to know about quantum physics is that it's all about probability—we can't say what will happen because there are so many possible answers. This is because, at the very heart of quantum physics, we deal with waves of probability as opposed to matter or energy, which exist in definite states such as beings or waves.

This chapter will go over some fundamental concepts in quantum physics, such as perspectives on reality, particles versus waves, uncertainty principles, and how they affect different levels of observation. We'll also discuss the mind-body problem and how quantum physics can be used to influence the world by altering the probabilities of what we perceive.

Quantum physics is a very complex, confusing, and controversial field of study. This article will attempt to cover some of the basics while providing links for further study in each section. For those interested in diving even deeper, I recommend starting with Quantum Mechanics: Theory and Experiment by Herbert P Feynman.

Perspectives on Reality

In quantum mechanics, our perception of reality is like a movie in which there are many things going on that we don't notice because they are outside the frame. These things are not real to us simply because we cannot perceive them directly. Within the context of the movie, quantum physics deals with three fundamental perspectives on reality: particles, objects, and observers.

Particles and Waves

In everyday life, our consciousness perceives the world in terms of objects that occupy localized positions in physical space. When light reflects off of an object, it forms what we call an image which is a matter-like object that can be measured and described by science. But when light interacts with very small particles such as atoms or photons (light particles), they become waves that can't be localized in a specific position; these waves are described by quantum theory instead of classical physics.

Quantum theory views matter as waves and light as matter; what's more, is that the laws that govern light waves are also applied to matter and vice versa. Here we see how quantum theory merges the concepts of matter and energy, which are normally viewed completely separately in classical physics.

Particles can be viewed in two ways: either as localized objects or waves of probability known as a "probability wave." When an object is observed, it is forced to choose one state or another, losing its wave-like properties in the process. This means that until an object is observed, it exists in a state of possibility known as a "superposition. When observed, the superposition collapses, and what was a probability wave becomes a localized particle. If you think about it this way, our observation is what causes things to happen in the world.

It is impossible to know both the position and momentum of an object at the same time (this is known as the uncertainty principle). An object's position can be noted but not its velocity, while its velocity can be measured but not its position. Another way of looking at this is to say that physical systems change over time, which means our perception of their state changes with them. This leads us to another concept called "decoherence," wherein quantum states decouple from their environment and become unavailable for future measurements. A system can be thought of as decohering when its properties "decide' which values they will have.

Objects vs. Observers

Before we can perceive the world around us, photons of light must interact with our retina and enter our brain via the optic nerve. Theoretically speaking, an object in the quantum world can only collapse into a specific state after being observed by a conscious observer. This means that without observation or measurement by a conscious observer, there is no reality; things only become real when we look at them. It is our consciousness that causes what were once probabilities (waves) to collapse into localized objects (matter).

Considerations of Observers

Observers themselves can be described as objects that exist in space and time and are composed of subatomic particles. Theoretically speaking, observers do not exist until they are observed by another conscious observer, meaning that without observation by a conscious observer, there is no reality. Likewise, all objects in existence must be observed by a conscious observer for the probability wave to collapse into a specific state; without this action, an object will continue to exist in a probabilistic superposition of states. This means that consciousness plays an active role in creating our reality and perception of the world around us.

Uncertainty Principles

The Uncertainty Principle is a concept that states it is impossible to exactly know the state of a quantum system. In other words, quantum theory says that we can never know both the position and velocity of an object at the same time; this is known as the uncertainty principle. This means that physical systems change over time, which means our perception of their state changes with them. The uncertainty principle plays a crucial role in how quantum physics is interpreted; it has been called into question by some but largely stands as one of the cornerstones in how quantum physics works.

Describing something as "a particle of light" is a poor choice of words since there is no such thing as a particle. In fact, light behaves like both matter and waves. For example, when they travel through space, photons travel in a straight line; but when they are passing through a medium, light refracts (bends) like a wave. At the quantum level, photons also behave like particles as well. This means that individual photons can be counted just like footballs or baseballs; you can know how many photons exist and where they are in an instant. Photons can also be described as an electromagnetic wave that vibrates at a certain frequency which we call their wavelength.

Photons exhibil both particle and wave properties depending on how they are observed. This is why they are referred to as "wavicles." Analogously speaking, when we throw a rock into the water, it behaves like a particle and scatters, but when we ripple the water with a paddle, it behaves like a wave, back and forth.

However, now imagine that I had never seen anything listed as "particle" before I learned about matter. I am told about the subatomic world and learn about "particles" made of even smaller things which are called "protons" and "neutrons." I am told that it is impossible to know both the position and velocity of this matter at the same time and then told that it doesn't exist in one location like a rock or baseball but instead exists in multiple locations at once. This is a concept that I could never understand until I was given a reason to believe in such nonsense. I think we need to answer some very basic questions about how the quantum theory works before we jump to conclusions on what it means about reality.

There is no such thing as 'particles' or 'solid objects.' Some physicists would prefer the term 'object' while others stick with 'particle,' which can lead to some confusion. Objects are made up of particles which are in turn made up of waves that behave like particles when observed. The word "particle" comes from the Latin word "particula," meaning small, and it is used to describe objects of very small size that exhibit a particle-like behavior. The word "object" came from the Latin termination-ōb, which means a thing or body. Objects can be described as something physical that exists in space and time, such as a rock or a baseball bat. An object is something that can be detected by the five senses. The word "quantum" comes from the Latin word "quantus" meaning how much or what amount and is used to describe fundamental units of matter. The term quantum alludes to the basic units of the matter being little packets of energy being released at the speed of light in Planck's constant. These are called quanta or photons. All those particles that make up objects travel at light-speed, and this is why there is no way to detect whether we're moving forward or backward.

Chapter 12

Law of Attraction: Introduction to the LOA

The Law of Attraction has become a term for a household. Clearly, it has become a buzzword for those learning ways to improve and enhance life. TV infomercials, movies, print media, and songs have become commonplace. However, the purpose of the law and its enforcement are two things. The literature is full of interpretations and explanations of the purpose of the law, but much too little attempt has been made to describe the physics of the law.

Much has been learned about the meaning of the law and how it can be applied, but the world is still waiting for the mechanics of the law to demonstrate how to really take advantage of it beyond simply keeping optimistic thoughts. What are the processes that make it work? Our efforts have centered on developing and designing resources that allow an individual to enforce the law more elegantly, effectively, and with less effort. What we discovered was a missing link on how to apply and execute this incredible fundamental theory of magnetism.

Research into quantum mechanics has shown that it is produced by the act of observing reality. Attempting to detect something allows nothing to appear. In the same way, if you don't know something, it doesn't exist in your subjective reality. In fact, the so-called placebo effect has shown that positive or negative behaviors will generate the corresponding effects. The work of Dossey and others has shown that prayer has an impact on whether or not the receiver is aware of it.

It is becoming increasingly evident that we are co-creating our reality in the way we think and feel... in other words, in our special and personal understanding of reality. We attract that which we consider as true into ourselves. This is the nature of the law of attraction that has been postulated and well-described in the literature since the New Thought Revolution at the turn of the 20th century all the way up to Esther and Jerry Hicks' current writings, and so appropriately exposed in recent movies like What the Bleep and The Hidden.

There is just a quantum of Energy as a stream of possible space positions and motions. This is omnipresent but, at the same time, none at all. There is no light and matter before something happens to make them "true," but what is that? Research has shown that this is something called the wave function collapse. And the further investigation has revealed that to manifest reality is consciousness, which performs this action.

The wave function includes all the potential outcomes of a given situation, but only one occurs in the physical world when a collapse of consciousness occurs. For now, scientists are all peering over each other's heads, watching the latest supercollider's event screen at Corn, Switzerland. They expect two protons to smash into each other, and when they do, the screen shows an "episode," showing the release of all the subatomic particles that make them up. What they do not understand is that their own screen observation causes the wave functions to appear, but because they have no intention of using them in any positive way, they simply vanish again. Realizing that each of our realities (body, mind, and spirit) is made of these energy particles whose wave functions have collapsed through our own observation, we may learn how to travel. Through witnessing what we want to manifest in our minds through imagination, inner listening, and a clear sense of our thoughts, we will learn to construct what we want. Likewise, we can resist the breakdown of possibilities into objects, events, and circumstances we don't want by refusing to give them our attention. A consciousness must experience a parcel of energy to be actual. Until that time, it lies enveloped on the other side of the quantum veil in the mystery of possibility. The "wave function" collapses when the energy is detected, making it measurable in the real world. Such particles can't be seen with the naked eye, but only through advanced devices that can reveal where the particle was, along with the speed and position it was detected at the moment.

Because both consciousness and spirit exist in the spiritual realm on the other side of the veil, it makes sense that when we are in an inspired state or good mood, we can control our reality even better. Once we are linked to the origin, we have more power in our own lives and in our entire sphere of influence to create real and constructive change. Once we remain in that state of grace and reverence, trends arise, synchronicity increases, and people who want to be around us and in that state of perfection want to join us. A positive wave function can more easily collapse while in this mode.

To stay in that state needs complete consciousness focus. The immediate knowledge of the divine, the one and all that leads to belief in the world that leads to the trust that is needed to step up to a higher-order level. We need to learn how to collapse only the positive, constructive wave functions while taking away our focus from those leading to negative, damaging effects. A single person's thought, speech, and behavior can alter the course of the global community as a whole: this phenomenon of "observing" reality often applies to the other sensory mechanisms. When you hear something from within your inner voice, the spirit of reality or higher self, or sense something deep inside your very heart, you do the same thing when you perceive it... you experience it. Experience is, therefore, a more apt word to characterize the true full existence of the operation of a wave collapse. And even better when the perception can be multi-sensory. Seeing, hearing, and go through an experience results in a state of sensory resonance, where all the sensory processes are coordinated, and a healthy autonomic nervous system contributes to a combined experiencing deep relaxation and immense creativity at the same time (see the theory of sensory resonance).

We each have a part that is animal and spiritual. We are not one of these things, but rather a consciousness sandwiched in between, whose task it is to choose between the options offered by these two opposing primary elements of ourselves and overcome the conflict generated by the discrepancy between the polarities: remember the old cartoons where a little angel appears on one side with a little devil on the other? It's Life.

When a person creates his or her own development and positive spiritual growth, he or she draws positive experience into him or herself. But when a person takes the wait and sees evolution approach, just sits around and "goes with the flow" by taking whatever is presented to them, they fall victim to the "strange attractor" or the chaos of nature that disassembles and demands rearrangement and reassembly to a higher level of order. The ancient Greeks, who believed he functioned as a ghost, called this cycle Klisis to "unravel" everything not considered to be in order.

Nature does not rule the world and our lives; they are inherently disordered... in chaos. Nevertheless, the four fundamental forces that physicists claim to materialize nature have recently been redefined as celestial attractors that, over time, establish patterns of order. That is the real mystery of how reality expresses itself. Space is the original force that creates the universe by a point or singularity of zero dimensions. The time is used to chain the points into the other attractors together.

An individual must become an attractor in order to make use of the Law of Attraction. Nevertheless, there are four attractor orders:

1. The attractor of the 1st order POINT – leads one to be drawn into one specific task or to get trapped in a rut by being too concentrated on one idea or fixation.

2. The 2nd order CYCLE attractor – allows one to get caught up in logical thought or an infinite loop that essentially repeats itself over and over again.

3. The TORI attractor 3rd order limit – which is a step in the right direction as it allows for a complex flow of Energy but is somewhat constrained due to its quasi-periodicity.

4. The 4th order ODD attractor-which is the absolute's chaotic behavior undoing everything that is not considered to be in order.

The emphasis, purpose, and consistency of action and expression that one pays and makes decide the attractor in which they fall. The best course of action is to align oneself with the Tori attractor, direct it and organize one's entire energy field in the constraints of that dynamic topological space, which limits it and allows it to become more periodic and repeatable, allowing the integration of chaotic change organized and controlled by an equal amount of determinism. In other words, a person may become an infinite toroid attractor, drawing into what they want instead of only taking whatever the strange chaotic attractor has to offer. This is the streamlined nature of the Attraction Law and the secret to its understanding and execution.

Looking at a spectral-domain image of a time-domain object through eyes-open meditation and observing such attractors in action that reveal secret desires or intentions beyond the threshold of consciousness. These pictures are deliberate palettes in which purpose can be visualized. All geometries and shapes have been shown to come from different ratios of real and imaginary partials that combine in the complex structure that makes up the wave function. Knowing this and that spoken or breathed speech is nothing more than a flow of these actual and imaginary elements, it becomes theoretically possible to learn how to draw what you want to see on the screen, hear what you want to hear on the headphones, feel what you want to feel inside, learn how to interact and direct the behavior of the automatons that make up all things.

The first step in learning to collapse the wave function is to perceive sense and experience what one thinks and feels on the inside. It, in turn, is the result of applying the law to attraction by studying how to connect with the ideal attractor while paying attention to the images on it—using a multi-sensory bi-phase approach which requires the user to become the agent of their own transformation by resolving the difference and tension between the operator and the machine, a polarity which can also be described as player vs. instrument, subjective vs. objective, actual vs. potential, or as the engineers know it as real vs. imaginary. These actual and imaginary components are called automatons and are the simplest, most indivisible elements or building blocks of existence. We are both operator and system at the most primal simple level of nature. Change at that point will allow us the ability to set up the next moment and permanently alter their lives. The epitome and final fate of the collapse of the wave mechanism are to achieve convergence between player and instrument.

For years psychologists have understood that visualizing a goal in an inkblot picture is the essence of combining several scenarios into the one that the viewer actually observes. An example is seeing inner-voice or music in white noise. The evidence of this trend is the incredible ring of reality that is always sensed in forms that cannot be readily described. Bringing those ideas to the quantum level and offering a means for an individual to imagine their purpose, listening to their inner voice, and witnessing their emotions at the most basic level of truth where life originates is one of our mission's key objectives.

Chapter 13

The Three Laws of the Universe

If you're curious about attraction, motion, and energy, you might be interested in some of the fundamental principles in quantum physics. The four laws of physics are not enough to describe all aspects of reality. Quantum Physics is needed for substances at a very small scale, like atoms or photons. There are three laws that every quantum physicist knows about:

1. Superposition Principle - The wave-like nature of matter implies that a particle is always in more than one state or place at once until it's observed or measured.

2. Observer Effect - The act itself of measuring something can significantly change its properties as well as our understanding and knowledge of what would have happened without the measurement being taken.

3. Wave-particle Duality - Matter can behave as either particles or waves.

Combined, they're called the Copenhagen Interpretation, where a particle is in many states, and a wave has an underlying particle nature to it.

For example, if you take a rock and drop it, the rock will fall to the ground. In Quantum Physics, you could explain why the rock falls by looking at each atom in table one at a time and seeing that when any atom comes close to an atom in the rock, it could form a bond with that atom and start drawing energy toward it and building up more potential energy between those two atoms than other atoms that are further away from that pair. So, although it's not a direct cause and effect relationship between the atom you're looking at and the rock, there is a cause and effect connection that's formed from the billions of atoms in both objects.

Bohr's model of the hydrogen atom can be visualized as a tiny solar system. The sun is like the proton at its center orbited by one orbiting electron. To understand why this magnet-like field attracts the orbiting electron like gravity attracts our moon to earth, this model requires Quantum Physics because it needs to describe both natures at its smallest scale, and it needs to describe all aspects of reality at its largest scale as well.

The electron is always moving, but it's also in orbit around the proton like you're always moving. But you're not really in orbit from moment to moment. You're just moving, but the earth is spinning faster than you're moving, so you can feel like you're in orbit. So, what we have is a particle-like aspect with an underlying wave-like aspect of electrons and protons as well as orbiting electrons, which are both particles or waves, depending on if it's being observed or not.

So, when you take a hydrogen atom and pass an electric current through the wire to the proton at the center of that atom in the shape of the solar system, it gets hot because hot electrons near cold particles repel each other. For example, if you have a tub of water with warm water near the top and cold water near the bottom, there's not much cohesive energy between the warm and cold water. But if you put a mix of warm and cold water into a spray bottle and shake it up, suddenly there's a lot more cohesive energy between the waves because they are at higher frequencies. If everything is hot all around you, like in an oven, that causes your body to go from being really hot towards being very cold because hot waves will naturally repel each other just like hot electrons will repel each other.

So, we have a proton with a positive charge at the center of an atom that's attracting an orbiting electron with its negative charge. If we increase the potential energy (like increasing the voltage) that wire is putting out, that attracts more electrons to orbit closer to this proton. At some point, if you increase it enough, it'll become unstable and either spit one or many electrons out of the atom-like shooting stars or fireballs. This can be hazardous to us and other things around us if they get hit with one of those free electrons flying through space at extremely high speeds when they leave the hydrogen atom because it's part of their momentum.

The wave-particle duality brings us to the Uncertainty Principle - the more you know about a particle, the less you know about its wave aspect and vice versa. This is because if you observe it with a microscope or any instrument, your observation will change its state from something that can behave as both waves or as particles.

The Uncertainty Principle is based on probability conservation. If we just look at tiny pieces of those objects without observing, information from different parts of those tiny pieces are undefined and can be anything we want because it's just probability.

If we observe a particle, it behaves as either waves or particles because the act of observation changes its state. As far as reality is concerned, it's only a wave if we leave it alone and don't measure it or observe it. Because of the Uncertainty Principle, there's nothing that can be said about its location because observing where the electron is will change its momentum.

For example, if you take two sheets of paper and put them underneath a microscope to look at them, you'll see something like this:

So, when I tell you that electrons are going to land in one place and not another, I'm only giving you extremely probable information about what could happen until you take an actual measurement.

That's why I say there's nothing in space because we're only giving you information about the probability of what could happen.

The basis of Quantum Physics is that anything that can happen will happen, but we don't know what will happen until an actual measurement is made. If we measure an electron on one sheet of paper and not the other, then it will definitely land on that piece of paper instead of the other one. But looking to see if your electron went to one or the other before you actually make a measurement didn't change its momentum as far as reality is concerned, so when you ask if it was going to hit either piece of paper, it was both pieces of paper.

As far as reality is concerned, the electron didn't land on either piece of paper until you actually went and looked to see if it landed there. So, for practical purposes, we can say that it lands on one side or the other with equal probability until we actually make a measurement.

But before we make that measurement, we have to consider everything else in the universe. If you're not alone in this universe (and you're not, so consider all those people around you), then they also have electrons flying around inside them with the same momentum and direction as yours.

All of those electrons are entangled with each other as well, so when you measure your electron going in one direction, the electrons that are in other people's bodies will also have their probabilities changed to reflect the fact that there's now twice as much momentum in that direction.

If we're giving you information about probability instead of actual reality, then it doesn't matter if it is a single particle or if it is an ocean full of those particles. You can't predict everything about them because reality is based on probability, not a certainty.

If I give you a ball and tell you it's got an even chance of being red or blue, that means its color will probably be red or blue before you open the box. If you gave me a ball like that and asked about its color, I could tell you with equal probability that it was red or blue without actually looking inside the box.

You can't predict which one it will be, but you can still answer any questions about the ball that don't rely on knowing its color.

The example above doesn't seem like a great one of quantum entanglement because we're talking about something we can see before we open the box, but that's just because our human brains evolved to understand a world where reality is based on our senses.

The visible spectrum is just a tiny part of all possible forms of light, and the rules of probability are just as true for light as they are for electrons.

Most of the time, it's too hard to measure our light experiments, but there are plenty of things we can do on the quantum scale.

One classic example, for instance, is using two particles with a known spin pointing in opposite directions (either both up or both down) and measuring which way they're spinning.

If we measure one particle's spin and find that it's downward-pointing, we know that the other particle must have an upward-pointing spin. This is also true if we measure the first particle and find that its spin is upward-pointing.

In other words, by measuring one particle, we can instantly determine the spin of the other particle, even though the two particles are very distant from each other and do not interact with each other except through the act of measurement.

Chapter 14

The Quantum Mind and Its Applications

Quantum physics is a rather new field of study that has opened up an interesting door to the workings of our universe. With quantum physics, we are able to see the world through the eyes of an atom and explore topics like quantum entanglement and quantum tunneling. It's not just theoretical exploration either; practical applications for this knowledge range from next-gen solar panels to a new understanding of how the memory works in our brain.

It's All About the Wave

One of the most common misconceptions about quantum theory is that it violates Einstein's Theory of Relativity by claiming that everything can be explained by something called a 'wave function.' A wave is, mathematically speaking, defined as a continuous and regular oscillation in a given direction. The obvious problem with this idea is that waves are physical objects. So how could this wave be something that potentially exists everywhere at once? The simple answer is—it can't. Instead, what we're dealing with here is actually nothing more than another way of describing the position of an object—one that takes into account both its position and momentum at any given moment.

That's not to say, however, that there aren't certain objects in our universe that exist in the form of wave-like functions. When scientists are studying a particle like an atom, they can see it behaving as either a wave or as a particle—but never both at once. This is because of something called wave-function collapse. To understand this concept, it might help to try and imagine what would happen if you were to take two mirrors and place them at opposite ends of a room. If you were then to stand in the middle of that room, you would see your reflection in both mirrors simultaneously. However, the light from your body would only show up on one of those mirrors at any given time—it would just appear that way.

Scientists believe that wave-function collapse occurs as soon as we start to measure an atom—which is weird when you think about it. The idea that something can be in more than one place at once is pretty strange, but our inability to perceive more than one thing at a time isn't nearly as unusual. So why do we have to pick? As far as science can tell right now, there just seems to be something inherently unstable about the nature of reality that makes us perceive things as either existing or not existing—never both at the same time.

Quantum Entanglement

Let's say you were to take a couple of electrons and place them up in the air, several feet apart from each other. What happens if one of those electrons is given a small push? If you were to only look at the one electron, there's a pretty good chance that it would move away from its starting point—but what if you also looked at the other electron? According to quantum theory, both electrons would respond at exactly the same time—even though they might be miles and miles apart. This phenomenon is known as quantum entanglement, and it has been confirmed by scientific research on dozens of occasions.

The most common example often used in scientific literature is with photons. If two photons are emitted at the same time from the same source, they will travel in opposite directions. If one of those photons is then measured by a scientist, the other photon will instantly change its behavior as soon as it is measured. This effect even works if the two photons are traveling in opposite directions—which means that they must be communicating with each other to some extent.

Many scientists are split on whether or not this phenomenon can be explained with classical physics or if it requires a quantum explanation, but there seems to be no question that it does, in fact, occur.

What is Quantum Tunneling?

As you might have already guessed, there's a lot of overlap between quantum entanglement and the idea of quantum tunneling. You see, the electrons that are involved in this phenomenon don't actually need to be in contact with each other to communicate; they can do it over vast distances—which leads us to the concept of quantum tunneling. This is actually a pretty interesting topic that has applications much further outside of quantum physics than one might expect. For example, if you've ever used a laser pointer, then you're familiar with this phenomenon because a laser pointer relies on photons jumping right through objects without actually interacting with them.

In order for something to do that, it really has to behave in much the same way as an electron in a closed shell would. Nuclear physicists have actually put this idea into practice in the real world, using a complex setup of mirrors and lasers to effectively create a "box" that electrons can jump out of. The implications for this type of technology go all the way up to new computer chips because quantum computers would require billions upon billions more quantum states than today's computers—making them millions (if not billions) of times faster at performing calculations.

Astonishingly, it's not just electrons that can act like this. Scientists have also proven that protons can tunnel through walls, which is why we now have PET (Positron Emission Tomography) scans. This technology was once only available in hospitals; with quantum tunneling, however, it can be shrunk down to something that might one day fit into your mobile phone – opening up a whole new world of possibilities when it comes to medical testing.

However, It Gets Weirder

That's just the tip of the iceberg when it comes to the weirdness of quantum physics, too. There are even more mind-bending theories out there about how particles behave at small scales—and all it takes is one big discovery to throw everything we think we know into chaos (as always). For example, in 1964, physicist John Bell came up with an idea called Bell's Theorem, which states that particles do indeed "talk" to each other—which is something that Einstein himself utterly rejected.

Even more mind-bendingly, a paper came out that described how it's possible for particles to instantly know about any and all changes made to them—even if they're on different sides of the universe. Again, this goes against pretty much everything we've come to understand about the universe—but it might also point toward the way we'll have to think in the future when it comes to quantum mechanics.

But What About the Large Scale?

The traditional view of quantum physics is that all of this weirdness disappears over large distances. That is, when we measure something on a large scale, it behaves just like we thought it did. However, this thinking began to change in the 1980s when scientists began to look at quantum mechanics at larger and larger scales.

Take the concept of superposition, for example, which says that particles can not only be in multiple places at once but that some particles can maintain multiple states until they are observed. This means that there's no real difference between a particle and a wave—which would explain why quantum particles behave so weirdly when they're measured. Yet, if you wait long enough, a particle will collapse into a single state—which brings us back to our world of classical physics.

Chapter 15

Future of Quantum Physics and Modern Problems to Solve

Quantum physics is the field of science that studies the behavior and impact of matter and its interactions with energy on a small scale, specifically phenomena where "atoms" or particles exist in a state that can only be defined as "a wave."

During my research, I became intrigued by how much amazing new work quantum physicists are doing. Although they may not have yet answered all the questions they are working to answer; there are many ways to get involved in this exciting new field. By reading this article, you too can learn about the future of quantum physics and modern problems we still need to solve.

Quantum Physics in the Future

As mentioned, quantum physicists are making new discoveries every day. They are tackling questions about how the universe works. There are still many unanswered questions; however, in the future, these will all be answered. Quantum physicists like to ask questions that may not seem relevant to what we do every day. However, they do have a lot of interesting implications. Many of these problems can even be applied to everyday life.

One quantum physicist, Dr. Lisa Randall of Harvard University, is trying to figure out what happens to all the antimatter that has ever existed in the universe. Other physicists are looking at how gravity works on a very small scale. Some physicists are trying to understand how couples can fall in love as well as why we are so biased toward certain things unless we can see both sides and are able to look at an issue from many different perspectives.

Currently, there is a focus on trying to understand how "quarks" (smaller particles) don't fade away when they have energy, and then they act as they do… which means these quarks have no mass. They are energetic. That is some of the most important work that is being done now. We also know that there are billions of galaxies in our universe and many undiscovered ones. We also have many new experiments we can do with quantum physics, and because of this, the future looks very exciting and bright for quantum physicists!

Quantum Physics in Modern Problems to Solve

Quantum physics has provided us with many tools to solve problems that we never thought were possible to solve before. For example, when a battery runs out, it doesn't matter how long we wait... it still won't charge. This is the thinking of classical physics. In quantum physics, there are many examples, such as electrons that don't act like they have mass—or they act in a strange way. This means that sometimes things can exist in so many states at once, and they will only choose one state at the last possible moment.

This means that we can use quantum physics to solve problems that people believed were unsolvable, even though we never thought of using these ideas before. For example, if you want to use a computer or cell phone faster and better, then you can use quantum behavior to do this. Of course, you will need to take your battery out of your cell phone and run it over with your car first, but it is possible... by the laws of quantum physics.

Quantum physicists are still trying to understand more about what goes on inside atoms. We know that there are a lot of particles, but we can't see them, so we can't investigate them in any way. We also know that when we study the smallest thing, we might find out something we didn't expect and discover something new about our world and universe that no one realized before.

With the new discoveries physicists have made, they have learned a lot about how electrons move through different materials or even within fluids. We have also created new technologies like the "quantum computer," which is a machine that uses quantum physics to handle information and more. There is a lot of exciting potential with these new discoveries.

The Bottom Line

Quantum physics is a new field that has only developed within the last 100 years. Until then, it was not known how much power and knowledge quantum physicists could unlock. Now we know there are exciting possibilities to unlock even more, expanding our understanding of the universe as well as applying this knowledge to solve problems in our everyday lives. Quantum physics is an exciting field with many unanswered questions and amazing future possibilities for discoveries!

Chapter 16

Quantum Super-Positioning

At whatever point you play the guitar and hear the agreement, you experience the waves' impacts. The hints of every arrangement consolidate as they arrive at your ear. On the outside of the lake, in the wake of tossing a little stone, something very similar occurs: the knobs meet and meet on their shoreline.

Sound waves and water waves are raised, complete by singular wave focuses framing another wave.

These two conditions have one shared factor: the rotating waves consolidate to coagulate their plummet. The outcome is a point-by-point proportion that delivers another wave.

Waves can depict molecules, electrons, and a few different occupants of the quantum universe. Yet, these waves don't demonstrate the development of people like water or wind.

Instead, their mobile pinnacles and valleys may have total qualities estimated by quantum resources, for example, position or force. The electron iota is showered into the orbital haze of chance.

For instance, the electrons circling a particle don't exist anyplace known to man, as does the Earth when it orbits the sun. Preferably, it is set in an orbital haze of chance. This space cloud is a practical 3D quantum wave comprised of mountains and valleys that change after some time and speak to the opportunity to get electrons in a given space.

The calculation of this wave shifts as per the quality of the electron. A surface can be made where two quantum waves—speaking to two degrees of electron vitality—are assembled, prompting another example of pinnacles and valleys. This changes where the electron is well on the way to be found, and the noticeable structures of the particle can be influenced.

Steve Rolston, President of the Department of Environment at the University of Maryland, clarifies why our everyday experience doesn't have a quantum scale.

It isn't unexpected to state that maybe an electron has two unique energies simultaneously or that it is in a few places simultaneously in this kind of amplification. If you consider an electron only a molecule, this won't be clear. In any case, when you think about an electron as an all-encompassing item, the overlay is more straightforward. Waves—including wave super-positions—are in numerous spots simultaneously.

The setting may appear to be oddly extraordinary now and again, for example, putting an apple close to orange and attempting to call it a banana; however, it is valid.

This is not, at this point, clear in standard quantum tests. Along these lines, a separate electron shaft (or another quantum molecule) is terminated into a layer containing two little sequential cones—the delicate identifier records when an electron hits the opposite side of the cuts.

At the point when electrons act like particles - consider little balls—so you can hope to see an example of two arrangements of locators with a set behind each space. At that point, the identifier follows the unsettling influence, as though every electron ventures like a wave between the two limits.

The wave goes through the two bumps simultaneously. The resulting aggravation makes numerous dull and splendid regions on the divider.

This is a test that requests feeling; however, quantum material science shows what occurs. These two parts express the substance, like the conduct of every person, by driving every one of them into a voice over the flap "past the left projection" and a "right lumbar" space ("right projection").

A Quantum object goes about as a wave and molecule. Slides are focused on singular particles, yet the following example is like that of a wave.

Chapter 17

The Black Body Radiation

Heated Bodies Radiate

For now, we will go to another puzzle that despises scientists as the new century turns (1900): how do warm corpses begin? There was a complete understanding of the system in question—the heat was known to cause particles and iotas to vibrate vigorously, and particles and molecules were proven examples of electrical charges. (Obviously, Newton was in good shape.) From the investigation of Hertz and others, Maxwell's prospects for light-emitting diversion cases have been confirmed in any case. It was known from Maxwell's circumstances that the radiation traveled at the speed of light, and in this case, it was understood that the light itself, along with the warm rays associated with the field, were actually electric waves. At the time, the picture was that when the body was heated, the subsequent vibrations on the subatomic and nuclear scale were inevitably removed—acknowledging at the time that Maxwell's idea of electromagnetism, which is the most efficient in the physical world, was legitimate and at the sub-atomic level, these desirable costs would have passed, perhaps emitting warmth and visible light.

How Is Radiation Absorbed?

What is meant by the expression "dark body"? The fact of the matter is that the radiation from the hot body depends in some way on the body being heated. To see this with great success, we should immediately support and consider how different materials store radiation. A few, such as glass, seem to bring light in any way—light passes directly. With a shiny metal surface, light is not included.

 It may be visible. Dark materials such as ash, light, and heat are completely compressed, and the equipment is warm.

How can we understand these various processes, such as light waves that adapt to changes in applications, making these charges affect and store energy from radiation? Thanks to the glass, it is clear that this is not happening, somehow very little. Why not? Full understanding of why it requires quantum equipment; however, the general idea is as follows: there are costs—electrons—in the glass that can fluctuate in the light of the affected electric field outside, but these charges are firmly attached to the molecules, and can only fluctuate in certain waves. (In quantum artisans, these charging vehicles occur when the electron moves from one circle to the next.) Noticeable, so there is no recurrence with a small wave, and from now on, the vigor is obtained. That's why glass is perfect for windows! Duh. However, the glass is not clear on certain waves outside the visible distance (as a rule, both infrared and light). These are waves where the distribution of electric charges on iotas or bonds can often fluctuate.

How can we understand the thinking of light through metal? A small metal has electrons allowed to travel at all power. This is what makes iron into metal: it conducts energy and heat effectively; both are actually transmitted by the flow of these transparent electrons. (All things considered, a little warmth is transferred to the vibrations.) But metals are irresistible because they sparkle—why is that? Once again, it is those free electrons: they are pressed into larger segments (compared to particles) by the electric field of the approaching light waves, and this stimulating flow comes from the electric field, much like the flow in a talking radio wire. The radiation is the reflected light. With a shiny metal surface, a slight glow of glamor is combined with the heat, it is automatically renewed, and that is, it is visible.

At the moment, what about looking at something that shines a light: there is no transmission and no display. We are approaching the best end with ashes.

Like steel, it will lead to the flow of electricity; however, not even a successful approach. There are unconnected electrons, which can travel at all energies but continue to hold objects - they have a short-term meaning. When they knocked, they caused a commotion, like the balls hitting the guards on a pinball machine, so they gave off a strong force in the heat. Apart from the fact that the electrons in the ash have a shorter duration compared to that of the noble metal, they are clearly compared to the electrons bound by iotas (as in glass) so that they can accelerate and gain energy in the electric field. In this way, they are powerful mediators in moving energy from light waves to heat.

Absorption and Excretion

After seeing how the ashes can get into the rays and transfer energy to the heat, shouldn't that be said about talking? For what reason does it transfer when it is heated?

The similarity of the pinball machine is still acceptable: think now of the pinball machine where the boundaries are, and so on. Vigorously vibrate because they are cared for vigorously. The (electron) balls he removes will be unexpectedly accelerated in every crash, and these acceleration charges propel electric waves. And, of course, metal electrons have long lines that are particularly long, the vibration of the shortcuts affects them the most, so they cannot function in a social event and transmit heat energy indefinitely. It is clear from such assumptions that large-scale radiation shields are the most acceptable manufacturers.

Indeed, we can be more accurate: the body emits rays at a given temperature and reappears just as it emits the same rays.

This has been demonstrated: the basic point is that if we think that a certain body can end up better than a transfer, then in a room full of things with the same temperature, it will add radiation from different bodies better than restoring energy to them. This means it will improve, and the rest of the room will be cold, rejecting the second law of thermodynamics. (We can use such a body to build a warm car that separates the filling as the room gets colder and colder!)

However, the metal shines when it is warm enough: why would it be so? As the temperature rises, the shortcut portion of the particles vibrates at a constant level; this movement disperses and speeds up electrons. Indeed, even glass is illuminated at temperatures high enough as electrons emit and move.

Two Basic Rules

The basic assumption is based on the radiation test view of the gap.

Stefan's Law (1879)

Total P power from one square meter of black area in temperature T travels as a fourth total temperature:

$$P = \sigma T4, \sigma = 5.67 \times 10{-8} \text{ watts / sq.m. / K4}$$

Five years later, in 1884, Boltzmann discovered this T4 behavior in theory: he used traditional thermodynamic thinking in a case loaded with electromagnetic radiation, using Maxwell's phenomena to associate the intensity and intensity of energy. (The minuscule ratio of the energy from the opening will obviously have a temperature-dependent similar to the radiation strength inside.) See what's going on in the notes on the subtlety of the decision.

Wien Relocation Act (1893)

As the temperature of the broiler shifts, so does the repetition where the radiation is transmitted more frequently. In fact, that is also legally related to the overall temperature: $f_{max} \propto T$

(Wien himself discovered this law by speculation in 1893, after Boltzmann's speculation about thermodynamic. It had recently been observed, wherever it is, equally, by the American astronomer Langley.)

Truth be told, this high rise in f_{max} and T is natural for everyone—when the metal is heated in a fire, the main visible rays (about 900K) are reddish, with very little visible and re-visible light. Further increase in T causes a darker shade from orange to yellow, eventually blue to higher temperatures (10,000K, or higher) when high radiation exposure is clearly visible.

This is a repetitive step where the greatest force is important in maintaining sun-related energy, for example, in kindergarten. The glass should let the sun's rays in; however, not allow the heat rays to come out. This is understandable on the grounds that the two rays are at a completely different frequency—5700K and, state, 300K—and there are direct-to-light objects that are inaccurate in infrared radiation. Kindergartens operate on the grounds that fmax changes with temperature.

Chapter 18

Evidence from Quantum Physics Supporting a Simulated Universe

Most physicists do not have much of a metaphysical streak, and they prefer not to wonder what all the means they research. Philosophizing doesn't pay the lease on them. Physics functions, but they just go for the flow; they just shut up and make estimates. On the other side, most theorists typically do not have the technical experience and context to wax lyrical and come to grips with contemporary physics. They don't want to advertise themselves too far out of the traditional model box of their establishment, whether physicists or philosophers. Typically, it is not considered a wise career step, particularly if you are on the upward and tenure academic path. As for me, I don't have to injure or ruin any academic career, so I'm going to mix physics and philosophy and think way, way, way beyond the traditional model frame. By appealing to the Artificial (Virtual Reality)

World situation, it would be 'explained' by quantum (particle) physics. The Nobel Prize committee, if I'm correct, knows where to find me!

Several macro-world mysteries can be 'explained' by resorting to a universe scenario in a simulated [virtual reality], from sculptures that wander (Easter Island) to the ideas of an afterlife to certain thoughts of déjà vu to memories of past lives to seed 'circles' to spirits, and so on and so on. Any of these anomalies, though, can be identified and discounted as belonging to the occult or pseudoscience. Anomalies from hardcore particle (quantum) physics, the most experimentally tested technology recognized and responsible for gismos and gadgets for about one-third of the global economy, are not so readily discounted. Despite many of the runs on the frame, Quantum Physics perspectives appear to be along the lines of.

Albert Einstein: God does not throw dice.

Niels Bohr: Anyone who is not shocked by the [quantum] theory hasn't understood it.

Richard Feynman: Nobody understands quantum physics.

And these are remarks made by renowned quantum physicists.

The secret to reality, in general, though, and yours, in particular, resides in the fundamentals (i.e. the Traditional Particle Model [Quantum] Physics) and how it is constructed from the ground up. These anomalies and how they can be explained are included in the fact. Time to dream of the impossible!

Causality

You appear to equate the loss of causality with free will on a macro scale. There is no prior explanation on what you want to have for dinner tonight, only a spur-of-the-moment impulse. All of itself is free will and free will, plain and easy.

You also find, though, that they also do as they please on a microscale of fields and forces and particles - there is little requirement for causality. Radioactivity offers a great illustration. There's no clear explanation of why one dysfunctional nucleus goes poof, and there's no similar copy residing next door. Actually, if anything like radioactivity occurs with no obvious purpose at all, but the phenomenon fits one particular mathematical relationship (one of several theoretical possibilities), then that definitely means some form of intelligent interference behind the scenes. For us to see, the text in Augmented Reality is on the screen.

You either have to agree that the fields, forces, and particles that collectively make up the Standard Particle Model (nee Quantum) Physics have free will and are thus somehow 'alive' and 'animated' in any way (although their free will come with some restrictions just as your free will come with restrictions) you can't flap your arms and escape or keep your breath underwater for three straight hours.

Missing in Action

An electron can have this sum of energy corresponding to this 'orbit' (around an atomic nucleus) or the energy level corresponding to another 'orbit' or that other energy level corresponding to a different 'orbit', etc., but not any energy level (and hence 'orbit') between-between (since energy comes in single indivisible quantum packets). Energy is also a discontinuous phenomenon; just like you may have several coins of five cents (here I'm talking about Australia) such as five cents, ten cents, fifteen cents, etc. You can't have a seven-cent coinage value or a nine-point-three-cent coinage value.

Here's the rub here. When an electron absorbs or lacks energy, the 'orbit' increases or decreases. So, where the heck is it as it grows or falls between 'orbits' that are allowable? Does it keep Alice Company in The Twilight Zone or another world or Wonderland? Being 'orbits' allowable in-between is equal to possessing a prohibited amount of energy that would correspond to that in-between condition. As you raise the amount of your pocket change from five cents to ten cents, it will be like getting a six, then seven, then eight, and then a nine-cent coin.

For the anomaly identified as quantum tunneling, a comparable condition may be had. There is a particle on one side of a barrier here, and there is a particle on the other side of the barrier—instantaneously—never to be noticed in between.

Of course, the program of virtual reality might quickly vanish and reappear as the quantum moved from one permissible 'orbit' to another permissible 'orbit' or as it tunneled through, under, or through the barrier.

All Things [Not] Transparent

In the center of Westminster Abbey, the traditional macro comparison of an atom (nucleus and electron field around it) tries to imagine a gnat (the nucleus) with a cloud of bacteria (the electrons) across the walls, ceiling, and floor and thereby surrounding the gnat. In other terms, within your regular molecule, there is a heck of a lot of vacant space. This may mean that without intersecting something, electromagnetic radiation, photons, would have no difficulty going straight on into an atom and thus becoming delayed on its uninterrupted path.

Then why isn't it all transparent? Why is light not passing straight into you? Why are some objects translucent (air, glass) and some things opaque with a certain wavelength/frequency of 'light' (light is not only visible light but infrared light and radio light and microwave light, etc.) and some stuff (of equal density and thickness) not because the energy content of 'power' photons can matter as it passes through 'empty' space for all practical purposes). In comparison, there are no electrical charging properties for photons that will impede their movement directly into the ordinary atom.

Light moves through the air—a swift, anomalous stage. You can see straight into an environment of around 100 miles to see the sun, moon, stars, etc. Attach a smog or fog, and things are a little on the opaque side, but it's not significantly different from the average thickness and density of the clean air or an air-smog combination. It's a 99.99% empty room still. Somewhere, everything is screwy unless there is extra programming that counters the case, of course.

Chapter 19
Quantum Entanglement and Teleportation

Quantum entanglement is indeed spooky—that's something we can't argue with. Truly, it often seems more like magic than science, but then, many parts of quantum physics wind up feeling this way. We still have much to learn and explain about the world, and quantum entanglement is among the most mysterious and least understood.

When two particles are entangled, it means that any change to one particle in the pair, no matter where that particle is in the universe, will result in compensation by the other particle. According to Einstein's theory of relativity, the light was the fastest thing in observable existence, so this bothered him immensely. If, for example, a change in one entangled atom was made on Mars, and the other atom on Earth instantly shifted to compensate, this would be many times faster than the speed of light.

To understand quantum entanglement, you must first be introduced to the concept of spin. For ease of understanding, let's assume that all particles in the universe are spinning up or down and that the universe must always have a net spin of zero (that means there must be an equal number of particles spinning upwards and an equal number spinning downwards). If, by chance, someone changes the spin of one of these particles to be the other direction, the other entangled particle, wherever it is in the universe, will switch directions to compensate. This becomes even weirder when we noted that when we measure the spin of a particle, it will change *simply because we measured it* (this harkens back to the Copenhagen interpretation). However, the spin of one entangled particle will *always* be the opposite of its partner particle. This means that we can essentially change the nature of a particle across the universe instantaneously, simply by measuring the spin of its entangled particle.

John Bell proved this irrevocably in 1964 when he introduced one of the greatest theories of quantum physics: Bell's Theorem. Einstein wanted to believe that these entangled particles, rather than relying on "spooky stuff," had hidden information incorporated into them during their creation that would determine the way they would spin when measured. However, Bell's theorem irrevocably proved this to be false using simple math. Without getting too deep into the math, Bell essentially tested out all the scenarios that could arise from these particles having "hidden information" within them and solved for the math that would result from these measurements. What he came up with was that, no matter what was done, no matter how many times he measured it, there was no way that the math would allow for hidden, local information within the particles; rather, quantum entanglement was the only acceptable solution to Bell's Theorem.

Interestingly enough for us, teleportation is theoretically possible through the use of quantum entanglement. Physicists have used, to date, quantum entanglement to teleport atoms, electrons, and photons instantaneously between one place and another—so, someday, teleportation may be possible for us, as well. Essentially, in order to teleport a particle via this process, it needs to be combined and entangled with an entangled particle in a specific way. Using a lot of complicated math, and essentially a process that would involve "measuring" the other particle in order to instantaneously teleport the extra particle, it is possible to do, and it has been done. However, it's extremely difficult, currently, with particles much bigger than an atom and only works across limited distances. For the moment, it's still not possible for us to teleport people—or living things, for that matter. It may be possible someday, though.

Chapter 20
Introduction to Scattering

Assessing states in the continuum is a way of handling the scattering theory. The structureless position unit of mass represented with (m) and occurrence momentum signified by () shall be considered from a permanent scattering point through an interaction explainable by a latent factor V (). It is the equivalent of a one-body issue for the comparative movement of both structureless position units of masses such as m1 and m2. This could lead to a reduced group denoted by m1 + m2 / (m1m2). However, this can interact through potential, with () being the comparative movement vector. This can become

Here are some generalizations to make for scattering processes in quantum mechanics:

a. The simple potential account of the scattering process may contain some shortcomings. This challenge could cause non-local potentials. Therefore, it may not be easy to make a simple likely description in the scattering of particles at relative energy.

b. If the interface involving the particles contains spin-dependent forces, the particles could have spun. This will cause the spin alignments to rearrange during the scattering processes.

c. The particles may be composites with various constituent particles and not treatable as point particles. Excitations of inner levels of freedom of the particles could have occurred during the process of collision. This is known as rearrangement collisions or inelastic collisions. In this condition, constituent elements from projectiles could be transmitted to the target, or elements from the target could be transferred to the pellets.

Theory of Classical Scattering

Classical scattering is also regarded as elastic scattering, coherent scattering, or unmodified scattering. It is among the three types of photon interaction taking place when the force of the gamma photon or x-ray is little in contrast to the ionization potential of the atom. This can happen with low-energy radiation.

The photon has not enough potential for freeing the electron from its bound state during interactions with the attenuating medium. However, the potential of the photon is lower than the binding force of the electron. Therefore, no transmission of power occurs and no deposition of energy too. Classical scattering involves an alteration of the course of these photons. That is the reason why it is even regarded as unmodified scattering. At normal energy levels, it isn't a critical interaction method, even during radiography.

The atomic figure of the absorber and incident photon energy makes classical scattering vary.

Theory of Quantum Scattering

This scattering theory is a framework for learning and comprehending the distribution of waves and particles. This aligns with the collision and dispersal of waves with some material object. For example, a rainbow is created by scattering sunlight by raindrops.

The interface of billiard balls on a table, change of angle of alpha particles using gold nuclei, scattering of fission traversing a thin foil, and diffraction or scattering of electrons and x-rays by a group of atoms are all scattering processes.

Scattering also involves studying how partial disparity equations propagate quickly, bound, and cooperates with one another, even with a borderline situation, and propagating into the future. The principal or direct problem of scattering is how to determine the circulation of dispersed particle flux or radiation, depending on the features of the scattering factor. Also, the inverse problem of dispersion is the challenge of influencing the elements of objects, such as their internal structure and shape, by measuring radiation data or substances dispersed from the point.

It is useful for radiolocation, as the issue can serve various applications. These include quantum field theory, medical imaging, echolocation, nondestructive testing, and geophysical survey.

The Differences Between Elastic Scattering and Inelastic Scattering

Elastic scattering holds that the interior structures of the dispersed particles never change and remain unaffected from the dispersion method. Also, inelastic scattering features particles with altered internal systems. This could excite the elements of a spreading atom, and it can destroy the scattering particle causing the formation of totally new substances.

The Analysis of Partial Wave

An important method for resolving issues in scattering by decomposing each wave into various individual angular parts of momentum and solving it using boundary situations in quantum mechanics is called partial wave analysis.

Another way of expressing this scattering theory is by using a stable and secure ray of particles that can be scattered away from a globular symmetric potential V(r). Therefore, long distances represented by $r \rightarrow$ could be short-ranged in a way the particles will work like liberated particles.

Strategy and Formalism in Scattering

We can express a brief and resourceful scattering matrix formalism that can enable us to make a stable analysis of elastic wave propagation in multiple layers of anisotropic solids. With this, you can resolve issues of numerical instability familiar with the transfer matrix model. It can use the delta operator method to prevent broad reformulation through its modified versions. Every scattering matrix is gotten instantly, and you don't have to use the scatterer operator concept, also called the wave-propagator concept. However, it is comparable to the previous reflection matrix formalisms.

The global and local reflection and transmission matrices align with scattering processes in several layers, which are derivable. How to derive global scattering matrices based on the local ones is done depending on physical arguments. This will provide a better insight into scattering techniques.

Eigensolutions are another formulation method considered more concise for getting the universal dispersion matrices straightaway. The algorithms and expressions involved are efficient, brief, and suitable for handling.

Born Approximation in Scattering

Born approximation involves the process of using the incident field for the whole area, and at each point in the scatterer, it serves as the driving field. At the beginning of the development of quantum theory, this approximation technique is coined after Max Born, the proponent.

An elongate body can use it as the disturbance method for scattering. When the dispersed area is little contrasted to the event field on the scatterer, the Born ballpark figure is accurate. However, we can prove this theory using the dispersion of radio waves through a luminous Styrofoam piece. By assuming that each part of the plastic is polarized through the same electric field present at that center without the column, you can approximate it. At this end, you can calculate or estimate the scattering over that polarization as an essential aspect of radiation.

The Phase Shifts or Phase Difference

When two or new alternating quantities reach their zero or maximum values, you can use the phase shift or phase difference to express and define the difference in radians or degrees. A sinusoidal waveform is an alternating quantity featured graphically in the instance domain all through a horizontal zero axis. Sine waves functioning as alternating quantity possess a maximum positive value at the time represented by /2. It also has a maximum negative charge at the moment 3/2 and zero costs positioned beside the baseline at 2 and 0.

From experience, all sinusoidal waveforms cannot go through the axis point labeled zero simultaneously. But, if contrasted to a different sine wave, they can be shifted either to both sides of the zero-axis point by other values. For instance, if we align the waveform of a voltage to the waveform of a voltage, there will be a phase disparity or angular shift between both sinusoidal waveforms. After this comparison, every sine wave that cannot go all the way through zero at t equals zero has encountered a phase alteration.

The phase transfer is regarded as a Sinusoidal waveform and represented with the Greek letter called Phi Φ. The angle is calculated in degrees (grades)or radians, and it is used for indicating that the waveform has transferred from a particular indication point through the parallel zero axes. Also, the tangential discrepancy existing between several waveforms along a regular axis having sinusoidal waveforms with similar frequency is regarded as a phase diversity or phase transfer.

Conclusion

Quantum Physics is hard to understand and often infuriatingly complex. However, it is a fascinating branch of science. It has wide-ranging applications such as medicine and astronomy, but its most important application is in technology. Quantum Computing promises to revolutionize our world, but it is still in its infancy.

One of the goals of Quantum Physics is to understand the potentials of computing in an entirely new way. This involves exploring aspects of time that already are very hard for us to witness.

Quantum Physics takes us "beyond" the four dimensions that we observe in our everyday universe. Our space-time universe has four dimensions—length, width, height, and time. However, the fabric of space-time is not solid like a 3D solid object such as a chair or a guitar. It's more like a liquid or a gas. The fabric of space-time can be stretched and folded in surprising ways. Opposing forces are needed to hold these forces together—just like you need opposing forces to stop a bike wheel from going off the road!

Quantum Physics also describes how energy can be "entangled" with time—like how an entangled twin phone creates a unified picture when they are put back together again. Entangled twins happen all the time where two objects become one object; however, entanglement occurs only when particles become one body (such as when two electrons become one atom).

A quantum computer will use all this information to solve problems that take today's computers many years to solve. New computing technologies will come before old ones are even obsolete!

This past year, the world has seen unprecedented progress in the field of quantum physics. But don't let this scare you. We have some simple tips and tricks for you to learn about quantum physics that will allow you to use it in your daily life in a way that will help you understand the world around you.

Quantum Probability Principle

What is probability?

In quantum mechanics, probability is the relative likelihood of an atom, molecule, or particle being in a particular state at any given moment in time. Quantum mechanics uses probabilities to describe how particles interact with each other and the environment around them. Because probabilities are advanced by a certain amount every time they are measured, probabilities cannot be measured directly. For example, an atom's probability of being in a particular state can only be accurately measured when it is observed. This can be used to calculate the atom's probability when it is not observed. But we cannot observe an atom's unmeasurable probability directly because doing so would require altering it by observation. So, probabilities are effectively calculated, using logic and math to infer what is "likely."

The probability of this event happening is 1/50000. The odds against this are 50000 to 1. So even if I gave you one random number between 1 and 50000, you still would have a 1/50000 chance of winning this bet. Besides, odds imply that if someone bets $10 that you will win one time out of every 100 times you bet, they will win $100 out of every 100 times you bet $10. So, the odds that we have put together for you today make it very unlikely that we will win this bet, so low odds like 1/50000 work great for us!

Quantum mechanics are fantastic. They allow us to explain some of the most amazing things in the physical world. But they're a bit difficult to understand on a day-to-day basis.

Here is a simple explanation that will help you understand some of the basics of quantum mechanics. Let's start with a thought experiment:

Robby the Robot enters a room and walks towards an object sitting on a table. He reaches out his hand and touches the object. Upon contact, he hears a computerized beep. Then, the robot's hand gives a little jolt, and the robot experiences a small amount of pain. The computerized beep continues until Robby pulls his hand away from the object.

Can you guess what happens next?

Our robotic friend repeats this experiment again and again with different objects until he can "feel" each object's electrical resistance. He discovers that some objects feel like electrical wires, while some are as hard as rocks. In fact, sometimes he feels no resistance at all when he touches an object!

Like Robby, all objects in our universe also have their own unique physical properties. It turns out that there is no single set of properties for all objects in our universe; instead, different objects have different sets of properties! These different sets of properties are sometimes called "wave functions," "quantum numbers," or "quanta." For instance, one electron has a wave function, which has a value for each possible outcome of the measurement. When we measure its position on an atom, we get 1/2 + 1/2 = 1/4. But when we measure its momentum, we get -1/2 + 1/2 = -1/4! That's because it has a different set of wave functions than other electrons!

The basic premise behind Quantum Mechanics is that reality is not a single, fixed entity but rather a confusing mosaic of numerous microscopic components (including atoms and sub-atomic particles). Reality can be described as a "blob" of fundamental properties, which are all interdependent and governed by laws that are not always completely understood or accepted by the scientific community. There is no definitive source for these laws since they are constantly changing, and there is no way to predict what these laws will be when the next expedition into outer space, or a new particle accelerator, is run.

The phenomenon known as Schrodinger's Cat illustrates this concept. In a special box, along with a cat (which is both dead and alive), there is also an apparatus that can be used to observe the cat in three different states: dead (alive), alive (dead), or neither alive nor dead (alive/dead). This apparatus also has a lever attached to it. If the lever is pulled, both the cat and the apparatus are destroyed; however, if the lever is not pulled, then it does nothing because it cannot affect anything that isn't there. The logic behind this experiment is that if you don't know what state the cat is in (either alive or dead), then you can't make any assumptions about what state it must be in after you pull the lever.

In this sense, everything in our universe can be described as either "alive" or "dead," or a combination thereof.

www.ingramcontent.com/pod-product-compliance
Lightning Source LLC
Chambersburg PA
CBHW051113050726
47592CB00002B/804